Statokinetic Reflexes in
Equilibrium and Movement

Statokinetic Reflexes in Equilibrium and Movement

Tadashi Fukuda

Translated by
Nobuya Ushio and Jin Okubo

Translated from the Japanese edition
UNDO TO HEIKO NO HANSHA SEIRI, 2nd ed.
(Igakushoin, 1981) by IGAKU-SHOIN, Ltd., Tokyo.

English translation © UNIVERSITY OF TOKYO PRESS, 1984
ISBN 4-13-068109-4 (UTP 36807)
ISBN 0-86008-343-6

All rights reserved. No part of this publication may be reproduced or transmitted in
any form or by any means, electronic or mechanical, including photocopying, recording,
or information storage and retrieval system, without permission in writing from
the publisher.

Printed in Japan

UNIVERSITY OF TOKYO PRESS

Translated from the Japanese original
UNDŌ TO HEIKŌ NO HANSHA SEIRI, 2nd ed.
published 1981 by IGAKU SHOIN Ltd., Tokyo

English translation © UNIVERSITY OF TOKYO PRESS, 1984
ISBN 4-13-068100-1 (UTP 69016)
ISBN 0-86008-343-8

Printed in Japan

Foreword

For the past two decades this extraordinary monograph by Professor
Fukuda has provided an illuminating guide for Japanese researchers in
the fields of basic and clinical neurophysiology of posture and movement. It represents Professor Fukuda's ingenious global approach to
the understanding of complex motor phenomena and as such shows
deep insight into their intrinsic mechanisms. Publication of the English
edition of this book is particularly timely, when rigorous efforts are
being made all over the world to bridge the gap between our knowledge
of neurons at the cellular level and their integrated behavior at higher
levels of central nervous system function. This English edition enables
us to share the inspiring ideas flowing from page to page with our colleagues worldwide, and no doubt it will provide an impetus to further
development of the field.

Masao Ito
Department of Physiology
Faculty of Medicine
University of Tokyo

Preface to the English Edition

This book, a comprehensive introduction to neurotology, was first published in Japan in 1957, when neurotology was still a new field of research that had just begun to be developed. The book earned a good reputation and was widely read; demand for it continued to be great even after it was out of print, and a second, revised edition was published in 1981. The publisher, Igaku Shoin, decided to keep the original version intact, as it was considered a classic of Japanese medical literature. Some subsequent research work was added as a supplement in the second edition. This second edition is being read not only by medical practitioners but also by those in the realm of athletic activities.

Although certain portions of this book were published at various times in Europe or the U.S.A., it has never been translated into English in its entirety. Publication in English was first urged some time ago, and has finally been realized through the efforts of many young Japanese physicians as well as of Igaku Shoin, publisher of the Japanese edition, and the University of Tokyo Press, publisher of this English edition. I would like to express my heartfelt appreciation to all of them.

As an introduction to the book, its central scientific findings are summarized below.

1) The so-called pathologic reflexes, such as the tonic neck reflex and the Babinski reflex, which have traditionally been considered to indicate the presence of neurologic lesions, are in fact important postural reflexes that form the basis of human posture and movement; this has been illustrated using various postures represented in sports, pictures, and sculptures.

2) Two new vestibular tests have been developed and are presented: the vertical writing test with eyes covered and the stepping test. Numerous findings were obtained using these tests in clinical and fundamental fields of vestibulolabyrinthine physiology.

3) The physiology of training—the process of improvement through practice—is objectively described by showing the establishment of a new, higher-order, postural reflex. Termed the kinetic reflex, it is produced in

both experimental animals and human subjects by repeated stimulation of the vestibulolabyrinthine and visual organs. The postural reflex before this repeated stimulation is termed static reflex. The postures brought about by these two reflexes, static and kinetic, are symmetrical. The title of this book, *Statokinetic Reflexes in Equilibrium and Movement*, originates in this finding.

4) The general presence of the unidirectionality of optic eye nystagmus in animals has been established. That the vestibulolabyrinthine system functions in such a way as to augment this optic unidirectionality has been experimentally proven. Other findings on optic vestibular coordination are presented.

Special gratitude is expressed to Drs. Nobuya Ushio, Jin Okubo, Tsutai Nagata, and Elizabeth Ichihara as well as to Etsuko Hamao and Masami Yamaguchi of the University of Tokyo Press, without whose assistance this translation could not have been published.

Spring 1984

Tadashi Fukuda
Professor Emeritus
Gifu University Medical School

Preface

It has often occurred to me that I became a medical doctor by mistake. Since I was skeptical about everything and contrary by nature, I was neither able to follow the medical courses without questioning, nor was I able to absorb all the knowledge that was crammed into my head at school. I passed the examinations with great difficulty.

I chose to study otorhinolaryngology for two reasons: first, it did not require a long term of study; and second, at that time, the practice of otorhinolaryngology promised a sizable sum of money. I had no desire whatsoever then to achieve great feats, to become a great scholar, or to search for the truth in medicine. I was prepared to become a smart and cagey doctor. I was called up for military service in 1937, and after roaming through China for two years, I took up the oriental epicureanism of the Chinese philosopher Lin Yu-t'ang as my own creed.

After I was released from military service in 1939, I returned to medical school without really knowing whether I wanted to study or not. I enrolled in the graduate school and waited for my professor to assign a thesis topic. I assumed I would complete the thesis with great ease and that getting a doctor's degree would not take much effort. I was certain the diploma would make a good decoration for my office in the future, when I opened my own clinic.

Dr. Teiji Hoshino was my professor. One day I was called to his room and given a topic for my doctoral dissertation: *The Study of Judo*, a long-pending research subject in our department. I still recall very well my surprise, which was so great that I almost ran out of the office. In retrospect, it seems that Dr. Hoshino had assumed, because of my demonstrations of judo at several parties during the school term, that I had taken judo lessons in high school. Confronted with such a difficult research topic, I felt like a cornered rat.

The Study of Judo is the consideration and discussion of the complex and subtle attitudes of judo which vary from moment to moment from the perspective of physiological reflex. A study of this sort is not, however, as predictable and established as the measurement of a judo wrestler's blood pressure, the observation of the varied shape of his ears, or a test of his tiredness in the light of a Donaggio's test. It is almost impossible to capture objectively with the naked eye the judo postures, which vary at almost unimaginable speed. Even if we are able to photograph or film the rapid movements, we have not yet found any key to apprehending those attitudes.

At that time, there were not many references available for the study of judo. *Körperstellung* (Body Attitude) by Heinrich Gustav Magnus was the only book that Dr. Hoshino gave me as a reference. In addition, he showed me the very latest medical periodical of that time, the *Annals of Otology, Rhinology and Laryngology*, in which he had underlined in red an article entitled "Waltzing Test" in the table of contents, and told me that the Americans were studying the attitudes of the waltz.

I had not heard of Magnus before. For the first time, I timidly opened the thick book with a yellow cover, not knowing what to expect, and found countless pictures of strange postures among animals. I did not see any connection between the study of judo and the animal pictures. "Waltz attitudes," I learned from reading the *Annals* article, were far from the stylish and elegant attitudes of waltzing. On the contrary, the "Waltzing Test" was a new method testing abnormal and awkward gaits, derived from the conventional gait test.

I started with *Körperstellung* and the "Waltzing Test." It was a rather lonely and peculiar research life. In fact, there were neither teachers I could ask for advice, nor any method to which I could cling in my study. My research life was a strong contrast to that of my colleagues, who began to experiment with a number of rabbits, dogs, cats, and guinea pigs as soon as they finished reading books and journals; I was left with Magnus's *Körperstellung* and was completely shut off from the lively atmosphere of research life. I read the book over and over again, only to find it more and more difficult to understand, because of my ignorance of the anatomy of the cerebral nervous system.

I also read closely and attentively the articles on movement in a number of physiology texts. The texts of those days characteristically presented muscle movements in strobe pictures which recorded the movements of vocal chords, as well as sketches of them. But the texts spared only a few pages for articles on movement and never presented an analysis of the mechanism of human movements. If you take a look at the old physiology texts, you will see that they deal with the standing-up and gait attitudes in terms of gravity and that the study of human gait is more or less based on the concept that a plate with its flat and wide base and low center of gravity is more stable than a wine bottle. This concept can be applied not only to human beings, but also to inanimate objects.

Of necessity, I found myself trying very hard to observe living phenomena with my own eyes. I discarded all that I had memorized from my courses, discarded the conventional methods of experimentation, and tried to take a closer look at nature and life in a naive and unsophisticated manner. A strange thing happened. Viewed in this manner, all the knowledge of medicine which I thought I had gotten rid of, everything I had learned from the work of my predecessors, now took on new and fresh meaning.

It was really a great and significant experience for me to retrace the methods of my predecessors, to see how they conducted this or that experiment. It was both challenging and stimulating to go back to the essential factors of the ideas and the experiments that led Magnus, who was a pharmacologist by trade, to the study of attitudinal reflexes. I was fascinated especially with Eugen Goldstein, the towering figure whose work on neurology I happened upon while poring over references. It became my secret pleasure to read Goldstein's articles in the old German weekly medical magazines.

While I was struck by the greatness of Goldstein's thinking, testing, and the significance of the development of his theory day after day for about a month, I noticed that my own thinking for the first time was becoming full-fledged, as though it were ready to challenge this giant. It often happened that in the library, with Goldstein's thesis spread open on the desk, I would cover the page with my hands and become absorbed in searching for a clue to God's providence.

One winter day in the library—I still clearly recall that hot, steamy room full of morning sunlight—I invented the blindfold vertical writing test. For the first time, I had found the wings with which I could fly myself. The mysterious doors before which I had paused in such utter helplessness were now opened one by one.

Professor Koichi Nishihata had pointed out that the active attitudes of a child catching cicadas, or of a child throwing a spear, or of a child doing a broad jump correspond to the reflex pattern that Magnus discovered after long study. Building on this observation, I examined the forms, not only of judo, but also of other sports and acknowledged that Magnus's theory of reflex patterns of active attitudes could be applied to a number of activities.

Magnus's reflex had been verified in animals with underdeveloped cerebra, and also in the underdeveloped or injured cerebra of humans. It was therefore imperative to verify that Magnus's attitudinal reflexes are latent in average healthy humans as well. The vertical writing test finally solved this problem: it showed that Magnus's neck reflex pattern clearly existed in the vertical writing of a healthy person who had been blindfolded.

The vertical writing test also revealed several other facts. I remember that I was surprised to learn that a man with stenosis of the auditory tube whose writing had deviated to one side started writing normally after his ear canals were widened by Eustachian catheterization.

New findings came in like a torrent. Let me give one important example. The vertical writing test taught us very clearly that there were two phases to a labyrinthine reflex, and thus expanded the framework of labyrinthine physiology which had been established by Robert Bárány. In other countries, people whose thinking was trapped within the framework of Bárány's theory developed peculiar theories and methods such as cupulometry and subliminal rotation. If it had been evident earlier that a labyrinthine reflex has two phases, such theories would not have appeared.

This book is compiled from the treatises I published over the past 18 years. I have attempted to compile all of the articles in their original forms without tampering with them and altering them because I felt it my responsibility to present them in the forms actually published. Each chapter is complete and coherent since it was originally published independently.

The number of figures has been kept to a minimum. Nevertheless, I would like to emphasize the fact that each chart, and even each sentence, is the result of many months' experimentation and effort.

In compiling my treatises into one volume, I began to evaluate my twenty years of work. Although I always glanced at foreign research periodicals, I was hardly impressed or disturbed by what I read. As a consequence, I was able to develop a field of study of which the European and American academic world is still ignorant, and which is completely different from Western research on reflex physiology, balance physiology, and labyrinthine physiology. While editing and correcting the articles collected here over the past year and a half, I sensed in them something unmistakably and characteristically Japanese. During the time I was doing my solitary research, Professor Hoshino perceived my solitude and turmoil very well. I would like, therefore, to dedicate this book to Professor Teiji Hoshino in gratitude for his instruction and his deep affection for me.

I had the good fortune to be constantly guided, helped and loved by a number of colleagues of the academic world from different schools. Dr. Kinji Satta, Dr. Kiichi Nishihata, Dr. Ko Hirasawa and other professors and friends gave me instruction, advice, and help. I would like to express my thanks to them.

The faces of the friends who helped me with these articles and the faces of my colleagues now come to mind. Even though I cannot list each person by name here, I am sure they will be glad that it is published.

This book is not my own accomplishment, but the result of effort by a number of people, with myself as a central figure. This research is a crystallization of effort by Dr. Manabi Hinoki, Dr. Takashi Tokita, Dr. Koji Oku and others. Although they were busy treating their patients and operating on them during the day, they worked very hard until midnight experimenting and testing. I bothered them a great deal with editing, correcting, and indexing this book. The book, therefore, should not be called "my" work, but should rather be called "our" book. I find in the collaboration of work with others a great joy and a pride as a human being. I would like to thank my beloved wife Ikuko who devoted herself to such a spoiled and selfish man as myself, and also Mr. Yukihiko Shimodori, my father-in-law, who drew almost all of the pictures in the book. Lastly, I would like to express my gratitude to Igaku Shoin Ltd., which published the book, and to its staff members who looked after me in many ways until I finally finished it.

February 20th, 1957

Tadashi Fukuda
at the cottage near the
Nagara river

Translators' Note

Professor Tadashi Fukuda's book *Statokinetic Reflexes in Equilibrium and Movement* was first published in 1957, and was reprinted in 1974. When we first encountered it the book was out of print, and worn library copies would circulate among the doctors in otorhinolaryngology departments of hospitals all over Japan. The book fascinated us all; we read it many times, and the concepts described in it still continue to stimulate us in our research activities.

Dr. J. B. Baron, Dr. P. M. Gagey, Dr. J. Meyer, and Miss J. C. Bessinetone suggested that we translate Professor Fukuda's book to contribute to the exchange of knowledge between different countries and cultures. Ten years have now passed since we received approval from Professor Fukuda to translate his book. We would like to thank him in particular for his cooperation, and apologize to him that this translation took so long to complete. We are especially indebted to Dr. T. Nagata and Dr. E. Ichihara for their devoted assistance.

In compiling the Bibliography, we confined ourselves for the most part to works in Western languages. To supplement the contents of the original book, we have appended reprints of several of the articles Professor Fukuda has published in English over the years; these contain details of the experiments through which he arrived at his striking conclusions.

We hope that this book will continue to be regarded like a Bible in the field of postural reflexes and body equilibrium.

Jin Okubo
Department of Otorhinolaryngology
School of Medicine
Tokyo Medical and Dental University

Nobuya Ushio
Department of Otorhinolaryngology
Obihiro Kosei Hospital

Contents

Part 1

Studies on Human Dynamic Postures

Chapter 1
Tonic Neck Reflex and Tonic Labyrinthine Reflex

1.1. Introduction

In his famous book *Körperstellung* published in 1924, R. Magnus described various kinds of postural reflexes he had observed. He found that changing the position of the head of a decerebrated animal in relation either to its trunk or to the horizontal induced characteristic reflex movements; that is, flexion and/or extension of the four limbs. The reflex induced by moving the head in relation to the trunk he called the tonic neck reflex, while that induced by moving the head in relation to the horizontal he called the tonic labyrinthine reflex. Then he showed that, although these reflexes were typically observed in decerebrated animals, they were also present in normal animals as basic patterns in normal movements and that both the tonic neck reflex and the tonic labyrinthine reflex were involved in the following movements: rabbits crouching or raising their heads, cats raising or lowering their heads toward food, and dogs and horses running. He also pointed out that other examples of movements involving both reflexes could be observed in various animals in zoos and even depicted the reflexes of a polar bear walking round a cage as examples of these patterns of movement. These examples show that such reflexes definitely participate in normal postures of animals.

Although Magnus observed the presence of the tonic neck and labyrinthine reflexes in various animals with healthy brains, he failed to find them in a higher form of mammals, namely monkeys, that were not decerebrated. In other words, although he clearly and consistently observed these reflexes in decerebrated or anesthetized monkeys, he was unable to observe them in normal untreated animals. He concluded, however, that normal monkeys probably have these reflexes, although they were not detected, perhaps because the methods used for their detection were too simple.

Magnus also demonstrated, by use of numerous illustrations and

3

photographs, that these postural reflexes were manifest in certain clinical situations that resemble decerebration, such as in human fetuses with immature brains, children with various cerebral impairments (e.g., hydrocephalus and cerebral hemorrhage), and adult patients with various brain diseases (e.g., hemiplegia, brain tumor, and hemiparesis due to a gunshot wound in the head).

He failed, however, to detect these reflexes in normal healthy adults, and suggested that his inability to do so might be due to some masking factors, such as the action of some region in the higher brain stem, an influence of optical origin, or the fact that human beings are quite different from other animals in that they stand on only two feet. Magnus thus only presumed the existence of the tonic neck and tonic labyrinthine reflexes in normal healthy adults; he did not demonstrate them (Magnus, 1924).

In the following sections, I will attempt to answer the question posed by Magnus in the 1920s. I will adopt Magnus's postulation that the typical postural reflexes found in decerebrated animals are in fact basic reflexes governing the daily movements of normal animals, and I will try to prove that these reflexes are also basic to postures and movements of human adults. Although postural reflexes may seem to be entirely irrelevant to normal movements of humans, I will demonstrate that they actually function in normal adults. Furthermore, I will show that an individual assuming various dynamic postures in daily movements is actually doing so on the basis of these reflexes, especially when he exerts maximal muscular strength. I will also show that "good form" in various sports follows the reflex patterns of Magnus.

1.2. Basic Postural Reflexes

Before presenting my own work, I will briefly describe the patterns of the postural reflexes. These were clarified by Magnus in experimental animals as well as in patients with brain diseases.

1.2.1. Tonic Neck Reflex

This reflex can be observed in an animal that has been decerebrated and deprived of both labyrinths. In such an animal, the distribution of muscular tone in each of the four limbs shows characteristic changes in accordance with definite laws when the position of the head is changed from normal. The reflex is classified as follows according to the movement of the head.

Rotation of the Head: When the head is rotated against the trunk around the longitudinal axis of the body and fixed at a certain position, there is an increase in the extensor tone of the two limbs on the side of the nose and a decrease in extensor tone and subsequent flexion of the limbs on the opposite side. The two limbs on the side of the nose, i.e., on the side of the jaw, are named the jaw limbs (*Kieferbein*), whereas the other two limbs on the side of the skull are called the skull limbs (*Schädelbein*). Magnus reported that this reflex constitutes a basis for the daily movements of normal animals. He also saw this reflex pattern in patients with a decerebration-like condition. Figure 1.1 shows a typical tonic neck reflex induced by head rotation in a case which I examined of infantile paralysis due to brain hemorrhage.

Deviation of the Head toward One Shoulder: When the head is turned sideways toward one shoulder, there is an increase in the extensor tone of the two limbs on the side of this shoulder as the pinna approaches it and a decrease in extensor tone and subsequent flexion of the limbs on the other side. Figure 1.2 shows this reflex manifested in the same child as in Figure 1.1. This reflex is often not as clearly manifested as that of head rotation. In human beings, the extension and flexion of the four limbs sometimes occur in the reverse mode.

Ventroflexion and Dorsiflexion of the Head: When the head is ventroflexed or dorsiflexed relative to the trunk, the four limbs show symmetrical change. In all animals except rabbits, dorsiflexion induces extension of the forelimbs and flexion of the hind limbs, whereas ventroflexion produces the opposite changes. In rabbits, dorsiflexion causes extension

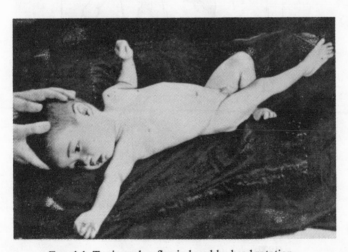

FIG. 1.1. Tonic neck reflex induced by head rotation.

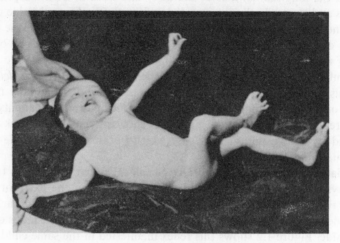

FIG. 1.2. Tonic neck reflex induced by head deviation.

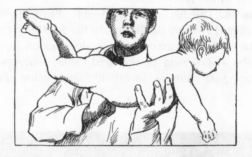

FIG. 1.3. Extension and flexion of four limbs induced by dorsi- and ventroflexion of head. Reproduced from Hoff and Schilder (1927).

of all four limbs, whereas ventroflexion causes flexion of all four limbs. In humans the pattern of reflexes is the same as in rabbits. Figure 1.3 shows these reflexes observed in a healthy infant. Extension of all four limbs accompanied by dorsiflexion of the trunk in the form of opisthotonus occurs on dorsiflexion of the head, while flexion of all four limbs with ventroflexion of the trunk occurs on ventroflexion of the head.

1.2.2. Tonic Labyrinthine Reflex

For demonstrating this reflex, the neck of a decerebrated cat is fixed in a plaster cast to immobilize it relative to the trunk. When this cat is turned around its frontal axis, its four limbs show change in extensor tone according to changes in the head position relative to the horizontal. The extensor tone is maximal, i.e., the four limbs are stretched most fully, when the head is in a position giving an angle of $+45°$ between the oral fissure and the horizontal, while it is minimal, i.e., the four limbs are bent, when the angle is $-135°$. There is thus a difference of exactly $180°$ around the bitemporal axis between the head position at which the tonic labyrinthine reflex acts maximally and that at which it acts minimally (Fig. 1.4). Magnus observed the same tonic labyrinthine reflex in infants with a decerebration-like condition, as well as in cases of amaurosis.

Thus one can say that the extensor tone of the four limbs in humans, as well as in cats, is maximal when the angle between the oral fissure and the horizon is $+45°$ and minimal when it is $-135°$.

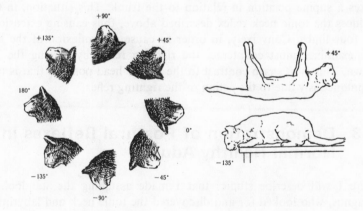

FIG. 1.4. Left: reproduced from Magnus (1924). Right: reproduced from Walsh (1957).

FIG. 1.5.   Righting reflex seen in a rabbit.

1.2.3. Righting Reflex

Every animal normally has a certain head and body position in relation to gravity. When the position of the head or the body changes in relation to the earth or the horizontal plane, reflex movements occur to restore the normal head or body position of the animal. This reflex is called the righting reflex. The vestibular organ is the most important in eliciting this reflex, but visual senses, and deep as well as superficial sensibility, are also involved. As shown in Figure 1.5, the head of a rabbit always assumes the normal position when the axis of its body is placed in a vertical position, irrespective of whether the head is above or below the body. This reflex may easily be observed in the daily movements of a healthy man, as will be described in detail later. As shown in Figure 1.3, when the body of an infant is forced to take a horizontal position, however, the head takes a nearly normal head position owing to the righting reflex. As the head recovers its normal position by means of the righting reflex, it consequently takes a supine position in relation to the trunk. This situation, in turn, induces the tonic neck reflex described above, thus causing extension of all four limbs. Conversely, in order to cause ventroflexion of the head, the examiner must counteract the righting reflex by forcing the head down. This is in sharp contrast to the supine head position that is taken spontaneously due to the action of the righting reflex.

1.3. Demonstration of Postural Reflexes in Normal Healthy Adults

Here I will describe studies that I made assuming the standpoint of Magnus, who looked for and discovered the tonic neck and labyrinthine reflexes functioning in the daily movements of normal animals. From this standpoint, I carefully examined various postures adopted by normal healthy adults in daily life and found that the postural reflexes, especially

FIG. 1.6.   Posture of a healthy adult catching a ball.

the tonic neck reflex, play very important roles in human postures.

It has been very frequently observed that when a man exerts his maximal muscular strength, he does so by unconsciously adopting the postural patterns described above. Figure 1.1 shows the tonic neck reflex observed in a case of infantile cerebral paralysis, a state of decerebration, and Figure 1.6 shows the posture of a normal healthy adult catching a ball. Comparing these two figures, we see that the two postures are identical in that the two limbs on the side of the nose, i.e., the jaw limbs, are extended to the full, while the two other extremities, i.e., the skull limbs, are fully bent. Obviously, this cannot be taken as proof that the posture of catching a ball is an expression of the tonic neck reflex in a normal healthy adult, because it is exhibited by a healthy adult without the slightest sign of decerebration, whereas the reflex posture is exhibited by a patient in a state of decerebration. Nevertheless, after discovering this fact, I thought that postural reflexes, especially the tonic neck reflex, might function latently in normal healthy adults: that is to say, although changes in the head position do not induce manifest extension or flexion of the arms and legs in normal healthy adults as they do in cases of cerebral impairment, these changes might induce unobservable extension or flexion of the limbs in the form of changes in the distribution of muscle tone.

Now, if it can be proved that the tonic neck reflex exists, although latently, in normal healthy adults, it may not be unreasonable to suppose that the posture of catching a ball described above occurs following the pattern of the tonic neck reflex, or that it is a manifestation of the tonic neck reflex in normal healthy adults. In other words, this posture may be induced by a volitional process, and the volition may intensify, facilitate, and manifest the tonic neck reflex, which otherwise would be acting latently, instead of inhibiting it, during the performance of any highly volitional movement, such as catching a ball. If so, it could be said that this posture is simply the result of a voluntary movement based on the tonic neck reflex. In consideration of this idea, I attempted to prove the latent existence of the tonic neck reflex in normal healthy adults and succeeded in demonstrating it in the following way.

1.3.1. Demonstration of the Tonic Neck Reflex by the Blindfold Vertical Writing Test

In 1959, I devised a test, the blindfold vertical writing test, in which subjects were blindfolded and asked to write vertically, which is the usual way in Japan (Fukuda, 1959a). Tests on many healthy adults with normal ears indicated that the tonic neck reflex did exist latently in normal healthy adults. Results showed that the arrangement of letters written while blindfolded and with the head in various positions showed either elongation, shortening, or deviation in a certain direction depending on the position of the head. These variations in the arrangement of letters on a piece of paper demonstrated that changes in the muscle tone of the arm took place in the same way as in the tonic neck reflex. This proved that the tonic neck reflex actually exists latently in normal healthy adults as far as the arm is concerned. This implies that the tonic neck reflex is not only a pathological finding in patients with cerebral lesions, but also a regulator, although latent, of relative movements of the head and the arm in healthy subjects.

Ventroflexion and Dorsiflexion of the Head: As shown in Figure 1.7, the series of letters written with the head ventroflexed is long, and that written with the head dorsiflexed is short. In each case, the subject intended to write the series of letters in the same way as those which he wrote when his head was in the normal position; however, he was unable to do so.

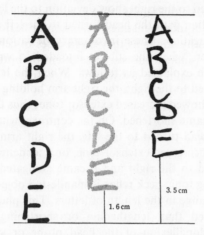

FIG. 1.7.   Series of letters written blindfolded with head ventroflexed (left), in the normal position (center), and dorsiflexed (right).

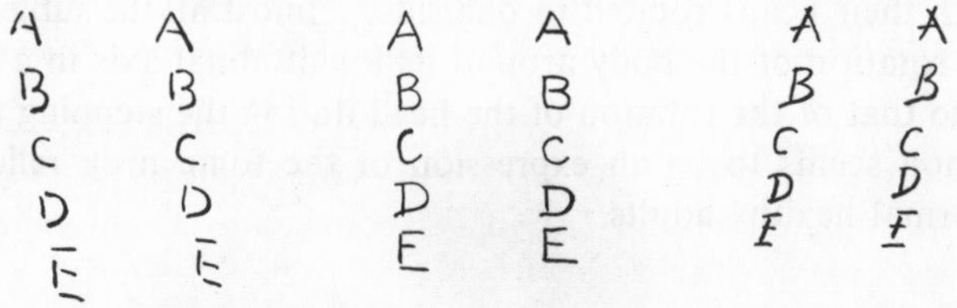

FIG. 1.8.   Series of letters written blindfolded with head rotated to the left (left), in the normal position (center), and rotated to the right (right).

This phenomenon can be interpreted as evidence for the latent existence of the tonic neck reflex, which exerts a bending influence on the arm holding the pencil upon ventroflexion of the head, and a stretching influence on it upon dorsiflexion. However, the tonic labyrinthine reflex also induces flexion of the limbs when the head is ventroflexed, and their extension when the head is dorsiflexed. Therefore, the above phenomenon can be considered to indicate the existence of the two reflexes, the tonic neck and labyrinthine reflexes, which are acting latently. This phenomenon has been observed in 70% of all the normal healthy adults tested.

Rotation toward the Right or Left: Figure 1.8 shows changes in the arrangement of letters induced by rotation of the head. The letters written with the head turned to the right show deviation to the left and shortening, whereas those written with the head turned to the left show deviation to the right and elongation. These involuntary deviations and changes in the length of letters when the subject intended to write vertically and consistently can be explained as follows. When the letters were written with the head turned to the right, the right arm holding the pencil became the jaw limb and showed increased extensor tone; thus the letters deviated to the left and became shortened. On the contrary, when the letters were written with the head turned to the left, the right arm became the skull limb and showed decreased extensor tone, or a tendency for flexion; thus the letters deviated to the right and became elongated. In this way, the latently functioning tonic neck reflex is manifested objectively in the form of deviation or change in the length of letters. This phenomenon was less frequently observed than lengthening or shortening of writing with ventroflexion or dorsiflexion of the head prone or supine, but it was observed in as many as 50–60% of the subjects we examined, the remainder giving irregular and inconsistent results. A similar tendency was observed when the subject inclined his head to the right or left, but the results were not so typical as those described above.

I also carried out the "stepping test" (Fukuda, 1959b) on normal adults with their heads rotated to one side. Almost all the subjects tested showed a rotation of the body around its longitudinal axis in a direction opposite to that of the rotation of the head during the stepping test. This phenomenon seems to be an expression of the tonic neck reflex in the legs of normal healthy adults.

1.3.2. Demonstration of the Tonic Neck and Labyrinthine Reflexes by Electromyography (EMG)

Tokizane and his associates (Tokizane *et al.*, 1951) demonstrated the presence of the tonic neck and labyrinthine reflexes in normal healthy adults by EMG studies. They placed needle electrodes in various extensor and flexor muscles of the four limbs and asked the subject to make slight voluntary muscular contractions so as to produce regular spike discharges in the EMG with almost constant inter-spike intervals. Changing the position of the head resulted in changes in the inter-spike intervals of a single motor unit in a manner indicating the presence of the tonic neck reflex. It is noteworthy that human adults were found to show extension of all four limbs in the supine head position and their flexion in the prone head position. This phenomenon in humans resembles that in rabbits, which, as mentioned above, manifest the tonic neck reflex in a way different from that in other animals.

1.4. Tonic Neck Reflex and Tonic Labyrinthine Reflex in Normal Healthy Adults

As is well known, skeletal muscles are also called voluntary muscles. Muscles of the neck and trunk and the four limbs can be voluntarily and freely bent or extended within the physiological limits of each joint. As shown in 1.3, the tonic neck and labyrinthine reflexes exert definite, but unobservable, influences on these muscles in healthy adults when the head position is changed relative to the trunk or to the horizontal plane. This indicates that the skeletal, or voluntary, muscles are not only controlled voluntarily, but are also subject to certain involuntary prescriptions, or movement patterns, which vary as the head position changes. What is the physiological significance of the involuntary movement of voluntary muscles? This question will be discussed next with special reference to the tonic neck reflex.

If one carefully studies a person catching a ball (Fig. 1.6), especially the position of the head relative to the trunk and the postures of the limbs, and if one compares this with Figure 1.1, showing the tonic neck reflex in a case of infantile paralysis, one is struck by the remarkable resemblance between the two. In both cases, the limbs on the occipital side, or skull limbs, are fully flexed, while the limbs on the side of the nose, or jaw limbs, are fully extended. If one interprets this resemblance to mean that the tonic neck reflex is working latently in the limbs of normal healthy adults, this posture of catching a ball can be regarded as a situation in which the latent tonic neck reflex becomes manifest; that is, the voluntary movement allows the tonic neck reflex to be manifest. This suggests that a man may be able to display his muscular power most efficiently if he assumes a posture conforming to this reflex. Now let us consider the physiological relationship between the voluntary control of muscles, or voluntary movement, and their involuntary control by means of the tonic neck reflex in the dynamic posture of catching a ball.

This posture can be thought to be composed of two kinds of movements: pure voluntary movement and unconscious reflex movement. The parts of the body on the right side of the line in Figure 1.6, i.e., the head, which is turned to the left, and the two limbs, which are stretched to the full, show movements produced by volition, whereas the flexion of the other two limbs on the left side of the line are involuntary. The player himself has no conscious intention of flexing these limbs or assuming this posture. His attention at this moment is concentrated on catching the ball. Thus, in order to reach a ball hit high above his left shoulder, full extension of the left arm and the left leg with a strong kicking movement of the left foot, which causes the body to jump off the ground, is produced by the player's will. In contrast, forward flexion of the right arm and backward flexion of the right leg are unconscious muscular movements.

As soon as the equilibrium of the body begins to break down owing to the voluntary movements of the body on the right side of the line in Figure 1.6, the tonic neck reflex is instantly and unconsciously brought into play to maintain equilibrium. When a ball is hit high above the left shoulder of the player who is about to catch it, his eyes follow the ball, and, accompanying this eye movement, the head is rotated to the left. Simultaneously, his left arm is stretched upward to catch the ball and his left leg is thrust down, causing the body to jump into the air. At this moment, his right arm is adducted at the shoulder and flexed at the elbow, and his right leg is slightly abducted at the hip and strongly flexed at the knee. Thus the equilibrium of the entire body is maintained on the stretched left leg, which is being used as a fulcrum. The flexion of the right limbs observed is simply flexion of the skull limbs induced by the tonic neck reflex, which is activated by turning the head to the left. In contrast, the extension of the left limbs is a voluntary movement. However, the tonic neck reflex provoked by turning the head to the left may intensify the extensor tone of the left limbs; the tonic neck reflex may intensify the voluntary movements so that the muscles can exert their maximum strength.

In summary, the tonic neck reflex plays the following roles in forming the posture of catching a ball:

a)　It promotes the extension of the left arm toward the ball so that the arm can be stretched maximally.

b)　It promotes the extension of the left leg which enables the body to jump into the air.

c)　It maintains the equilibrium of the entire body on the left leg, which serves as a fulcrum.

d)　It regulates the relationship between the head, trunk, and limbs.

An example of a similar physiological mechanism may be seen in Figure 1.9, which shows the posture adopted in "heading" a ball, that is, hitting a ball with the head. In this figure, the head is rotated to the left and inclined toward the left shoulder. The player's entire attention is directed to the movement of the head aimed at the ball. Movements of the other parts of the body are produced unconsciously by the action of the tonic neck reflex. The left arm is stretched fully at the elbow joint; the left leg is also extended; the right arm is bent fully at the elbow; the hand is flexed at the wrist, and the right leg is bent at the knee joint. This combination of flexion and extension of the four limbs coincides

FIG. 1.9.   Posture adopted in heading a ball.

quite well with the pattern of the tonic neck reflex induced when the head is rotated and inclined. In other words, these movements of the limbs are an involuntary reflex and are induced by the tonic neck reflex that is activated by the voluntary movement of the head. If the portions of the body assuming voluntary movements and those assuming involuntary, reflex movements are divided by a line, as in Figure 1.6, this line passes through the neck, as shown in the figure. Above this line movements are purely voluntary, and below it, movements are performed unconsciously and reflectively, and yet with maximal contraction of the muscles, especially in the four limbs.

FIG. 1.10. Judo throw using a foot trick.

Another example indicates how ingeniously the tonic neck reflex is utilized in judo. In Figure 1.10 the athlete on the left is challenging his opponent with a foot trick, *uchimata*. At the moment of the action shown in this figure, the challenger must bend his left leg maximally at the knee joint and stretch his right arm fully at the elbow and hand joints while holding the left sleeve of his opponent, in order to throw him. By pulling his opponent's body down and to the right, he brings the upper half of the opponent's body downward and the lower half upward, thereby attaining his objective. Of special interest in this posture is the position of his head relative to his trunk. At this instant his head is turned fully to the right. This rotation activates the tonic neck reflex and promotes extension of the right limbs and flexion of the left, thus unconsciously reinforcing the voluntary flexions and extensions of these limbs that are necessary for the maximal muscular force. Were the challenger's head turned to the left, this trick would surely fail to be effective. This example clearly shows how the tonic neck reflex is ingeniously utilized in human dynamic movements; that is, how the position of the head plays an important role in the bending and stretching of the arms and legs.

As exemplified above, the tonic neck reflex, which is provoked by rotation of the head and consists of extension of the jaw limbs and flexion of

FIG. 1.11

FIG. 1.12

FIG. 1.13

the skull limbs, forms a basic pattern of human movements. Further examples are given in Figures 1.11–1.13. These figures show the postures assumed at the moment of exertion of maximal muscular strength in various sports, such as archery, shot-put and high jump: the posture assumed at the moment of full extension or flexion is based on the tonic neck reflex. The relationship between the position of the head and the movements of the limbs, that is, extension and/or flexion, which is maintained while dancing or performing various gymnastics (Figs. 1.14–1.17) is also based on the same reflex. Of further interest is the fact that these

FIG. 1.14

FIG. 1.15   FIG. 1.16

FIG. 1.17   FIG. 1.18

natural postures, which are simply manifestations of the tonic neck reflex, are often portrayed by mannequins (Fig. 1.18).

The above examples are of asymmetrical movements of the four limbs induced by the tonic neck reflex with the head rotated. Next, let us consider symmetrical movements of the four limbs induced by changes in the head position relative to the trunk and to space. Figures 1.19 and 1.20 show two postures adopted while the body is in the air. In Figure 1.19, the head is bent sharply backward, the trunk dorsiflexed, and the four limbs maximally extended; in Figure 1.20, the head is bent sharply forward with the chin resting on the chest, the trunk ventroflexed, and the four limbs flexed. In these two figures, the relationships between the position of the head and

FIG. 1.19   FIG. 1.20

the trunk and between the flexion and/or extension of the four limbs and the ventro- or dorsiflexion of the trunk conform exactly to the tonic neck reflex that governs the postures of the limbs and the trunk when the head is ventroflexed or dorsiflexed as shown in Figure 1.3. In sports, coaches often recommend that their players allow their bodies to relax and move as naturally as possible. When this recommendation is followed, the player's body can be subjected to the involuntary action of the reflexes described above, thus producing the maximal muscular coordination. In other words, in sports a posture is beautiful and ideal when it coincides with the posture produced by the action of postural reflexes. For each position of the head relative to the trunk or to space there is a characteristic posture.

Figures 1.21 and 1.22 show two positions taken in a series of gymnastic exercises commonly practiced in Japan. As shown in Figure 1.21, the head is markedly dorsiflexed when the four limbs and trunk are stretched: that is, when the hip, knee, elbow, and finger joints are stretched, the shoulders abducted, and the trunk dorsiflexed. In contrast, the head is markedly ventroflexed when the four limbs and trunk are bent: that is, when the hip, knee, elbow, and finger joints are flexed, the shoulders adducted, and the trunk ventroflexed. In these two figures, the relationship between the position of the head and the flexion and/or extension of the four limbs also coincides exactly with the pattern of the tonic neck reflex appearing when the head is dorsiflexed or ventroflexed.

Furthermore, examination of the position of the head in these figures shows that in Figure 1.21 the angle between the oral fissure and the horizontal (measured according to Magnus's method, cf. Figure 1.4) is about $+45°$, at which the extensor tone is maximum in the tonic labyrinthine reflex; whereas in Figure 1.22, it is about $-135°$, at which the extensor tone is minimum. Thus it is seen that the extension of the four limbs in the former and their flexion in the latter are both intensified not only by the tonic neck reflex, but also by the tonic labyrinthine reflex. Thus the tonic neck reflex and the tonic labyrinthine reflex augment each other in intensifying the extension of the four limbs in the former posture and their flexion in the latter. In other words, these movements in gymnastic exercises are simply repeated voluntary movements constituted entirely by postural reflexes, i.e., the tonic neck reflex and the tonic labyrinthine reflex.

Now let us examine the posture assumed when standing on the hands as shown in Figure 1.23. In this posture, the arms are fully extended at the elbows, the entire weight of the body being supported on the arms and hands. The legs are extended at the hip, knee, and foot joints, and dorsiflexion of the trunk is observed. It is of particular interest that in

FIG. 1.21

FIG. 1.22

this position, the head is bent as far backward as it will go relative to the
trunk, and, by virtue of the tonic neck reflex thus induced, the four
limbs are fully extended and the trunk is dorsiflexed. Full extension of

FIG. 1.23

both arms at the elbows is a requisite for supporting the full weight of the head, trunk, and legs. The increase in extensor tone of the arms, caused by dorsiflexion of the head away from the trunk, is most effective for this purpose. On the other hand, the head is at an angle of about $-135°$, a position at which the extensor tone of the limbs becomes minimal. Thus in this posture, the tonic neck reflex increases the extensor tone, whereas the tonic labyrinthine reflex decreases the extensor tone. Therefore, the two reflexes act in opposition, producing the appropriate distribution of muscle tone throughout the entire body.

These opposite effects may explain why it is difficult to maintain the posture of standing on the hands for a long period. The posture assumed while standing on the hands is a clear example of one of the many human postures in which the tonic neck reflex plays a very important role.

1.5. Righting Reflexes

At the beginning of this chapter, I gave an example of a posture involving the righting reflex: the head of a rabbit always assumes the normal position relative to the horizontal, irrespective of the position of the trunk (Fig. 1.5).

FIG. 1.24

This righting reflex, shown to be provoked mainly by impulses from the
vestibular and visual, deep and superficial sensory systems, can readily be
observed in daily postures of normal healthy adults.

As shown above, in the posture of standing on the hands, the head is
dorsiflexed relative to the trunk so as to increase the extensor tone of the
four limbs by means of the tonic neck reflex (Fig. 1.23). In this instance,
the dorsiflexion of the head is produced by the righting reflex; that is, by
an attempt to maintain the normal head position while standing on the
hands. This head position happens to be optimal for inducing the tonic
neck reflex, which increases the extensor tone of the four limbs.

Figure 1.24 shows the posture of a girl walking on a horizontal bar.
If the body inclines to one side and begins to fall over, the head immediately moves to assume a normal position relative to the horizontal,
which consequently enables the trunk to return to about the normal,
vertical position. A similar example can be found in sumo, Japanese
wrestling. When one of the wrestlers at the margin of the wrestling arena

FIG. 1.25

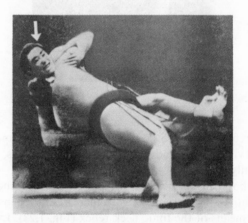

FIG. 1.26

tries to overcome the pushing force of his opponent, his body is tilted backwards, but his head attempts to maintain the normal vertical position relative to the horizontal plane (Fig. 1.25). The same situation is seen in Figure 1.26 where a wrestler is trying to counteract a lateral pushing force. Figure 1.27 shows *osaekomi*, a holding technique in judo. The trunk of the athlete at the bottom, who is trying to overcome his opponent's efforts to keep him down, is twisted as much as possible in an attempt to return to the normal position. Usually, in the righting reflex, the normal position

FIG. 1.27

is regained first by the head and then by the trunk. But in this situation, the head is immobilized by the opponent's abdomen, and, thus, it cannot be restored to the normal position, and the righting reflex is manifested by the trunk being twisted as much as possible in an effort to regain its normal position. In this posture, Magnus's *Körperstellreflex auf dem Körper* is manifested. Thus the righting reflex is always in action to maintain the normal position of either the head or the body.

1.6. Dynamic Postures Inconsistent with the Tonic Neck Reflex

As shown above, the tonic neck reflex forms a basis for various dynamic postures in normal healthy adults. In this section, postures that are not consistent with the tonic neck reflex will be discussed, and their causes or effects will be evaluated from a physiological viewpoint.

1.6.1. Pre-existing Accelerated Body Movement

Figure 1.28 shows postures in fencing. The relationship between the head and the arms is in good accordance with the tonic neck reflex as far as head rotation is concerned. The arm holding the sword, the jaw limb, is stretched, while the other arm, the skull limb, is bent. In this posture each fencer looks for an opportunity to attack his opponent. As soon as one of them finds this opportunity, he attempts to thrust his sword into the body of his opponent using as much muscular force as he can. At this moment, the body is moved forward very quickly, and the skull limb, which was bent before, is subconsciously extended with force (Fig. 1.29). Thus in this posture both the jaw limb and the skull limb are fully extended in association with rotation of the head against the trunk, a posture not consistent with the tonic neck reflex.

Fig. 1.28

Fig. 1.29

FIG. 1.30   FIG. 1.31

A similar situation is observed in tennis. When a player returns a ball from the place at which he is standing, no rapid lateral movement of the body is necessary. In this case the jaw limb is extended and the skull limb flexed in a posture typical of the tonic neck reflex (Fig. 1.30). On the other hand, when the player must move rapidly forward and to the left to reach the ball, that is, when additional lateral movement is necessary, both the skull limb and the jaw limb are fully extended, as shown in Figure 1.31.

Another example is seen in the postures adopted in catching a ball. When a man tries to catch a ball high above his shoulder, without having to move sideways, his posture is in good accordance with the tonic neck reflex, as described previously (Fig. 1.6, Fig. 1.32a). However, when the ball is further to the left, the posture he adopts for catching it (Fig. 1.32b) includes rapid movement of the body to the left. His right arm, or "the theoretical skull limb," which was bent in Figure 1.32a, is now stretched, conflicting with the posture of the tonic neck reflex. The right leg, however, is still bent and remains "the true skull limb." When the ball is hit still further to the left and is higher, even more rapid and greater movement to the left is necessary than that in Figure 1.32b, and the posture of the catcher (Fig. 1.32c) shows, besides maximal extension of the jaw limbs, maximal extension of the right arm and leg, which are the skull limbs of the tonic neck reflex, thereby apparently not conforming to the typical posture of the tonic neck reflex.

a   b   c

FIG. 1.32

These examples observed in fencing, tennis, and baseball are seemingly not consistent with the rule of the tonic neck reflex. However, careful examination shows that they are actually in accord with it. In these examples, extension of the skull limb is induced by addition of forward or lateral movement of the body. The physiological mechanism of the extension of the skull limb has not yet been clarified experimentally, but probably it is mainly induced by a reflex of vestibular origin provoked by a prompt forward or lateral movement of the body.

1.6.2. An Exception to the Rule of the Tonic Neck Reflex

Why is it difficult and awkward to use the backhand stroke in tennis? Why is this technique more difficult to master than the forehand stroke? I analyzed this problem from the viewpoint of the tonic neck reflex. When a right-handed person uses a backhand stroke, his right arm must be bent to receive and hit an approaching ball; that is, his arm must be bent at the elbow joint and adducted at the shoulder, and his forearm must be pronated (Fig. 1.33). At this moment, his eyes must follow the approaching ball, and so his head is rotated to the right relative to his trunk. Thus his right arm becomes the jaw limb and must show increased extensor tone. In other words, his right arm is bent by his conscious will, even though the tonic neck reflex prescribes stretching. In throwing a ball or putting a shot, the arm holding the ball or the shot is first bent (skull limb) and then stretched (jaw limb) at the moment of throwing the ball

FIG. 1.33   FIG. 1.34

or tossing. Thus the maximal muscular force is displayed when volition and the tonic neck reflex augment each other in flexion and extension of the limbs. In tennis a similar situation is seen during serving (Fig. 1.34) and forehand strokes; that is, the arm holding the racket faithfully follows the rule of the tonic neck reflex both before and during the strokes. In the backhand stroke, the tonic neck reflex is elicited at the moment the ball is hit (Fig. 1.30), whereas while the player is preparing to hit the ball, his will and the tonic neck reflex are seemingly in conflict with each other with regard to the bending and stretching of the arm holding the racket (Fig. 1.33). In other words, volitional flexion of the arm is counteracted by an increase in the tone of the extensor muscles caused by the tonic neck reflex, and so exertion of maximal flexion during the stroke is very hard. This is why it is so hard to execute a backhand stroke skillfully.

1.7. Human Dynamic Postures and Reflexes

Magnus could find no evidence for participation of the tonic neck and labyrinthine reflexes in daily movements of normal healthy adults. But, as described in the previous sections, I found that if the dynamic postures of normal healthy adults are examined carefully, they are found to be postures formed on the basis of these reflexes. Athletes participating in various sports assume postures in accordance with the patterns of the tonic neck and labyrinthine reflexes at moments when they have to exert their maximal muscular strength.

Magnus examined the usual postures of various normal animals and found that they were based on the tonic neck and labyrinthine reflexes that he had first observed experimentally in decerebrated animals and those deprived of labyrinthine organs. If these reflexes exist in normal animals, it seemed highly likely to me that they would also exist in normal healthy adults. Therefore, feeling certain that the tonic neck and labyrinthine reflexes were present latently in normal healthy adults, I tried to demonstrate them. My attempts were successful, and I could demonstrate the presence of these reflexes objectively by means of the writing test and the stepping test in most, although not all, of the healthy subjects I examined. The results of these tests showed that the postures were achieved by a combination of voluntary movements and movements caused by reflexes that are latently present in the human body, namely the tonic neck and labyrinthine reflexes. In short, these postures are manifestations of the tonic neck and labyrinthine reflexes themselves.

These reflexes have long been considered to be pathological, since they are observed only in decerebration-like conditions, and thus they have been the object of neurological examination only. However, the manifestation of these reflexes under pathological conditions is exceptional; they actually constitute the basic patterns of both static and dynamic normal human postures. I think that postural reflexes such as the tonic neck and labyrinthine reflexes are involved in the human extrapyramidal system. Their manifestation in normal healthy adults is usually inhibited by impulses arising from the cortex, the higher portion of the brain stem, etc. When maximal muscular strength is required, however, the pyramidal system actively combines with the extrapyramidal system, and thus these reflexes become manifest.

In reviewing the literature on this subject, I found that many people had attempted to demonstrate the latent presence of the tonic neck and labyrinthine reflexes in normal healthy adults. Investigators such as Wodak and Fischer (1923), Hoff and Schilder (1927), Luhan (1932), Pacella and Barrera (1940), and Gesell and Amatruda (1947) demonstrated the tonic neck and labyrinthine reflexes in newborn babies and infants by changes in posture, and proved their latent presence in adults by various procedures. Walsh (1957), citing Luhan's work, writes that "As time goes by, the responses of the musculature become more complex and in the adults the effects of the neck reflexes often pass unnoticed." Thus the demonstrations by various investigators using a variety of methods of the presence of the tonic neck and labyrinthine reflexes in normal adults lends strong support to my opinion. Furthermore, Nishihata (1938) and Hellebrandt (1956), taking examples from postures in daily movements and in several sports, pointed out that the postural reflexes exhibited peculiar traits and characteristics in normal adults.

1.7.1. Tonic Neck and Labyrinthine Reflexes in Art

It is surprising to find that artists of ancient times perceived the tonic neck reflex through their extremely keen powers of observation and depicted it in their works. A. Güttich (1942) carefully examined many pieces of old Greek sculpture and discovered that the posture of the tonic neck reflex underlay the postures of many of the statues. Giving examples, he pointed out that when the leg, the jaw limb, was extended in various static postures, the arm on the opposite side, the skull limb, tended to be flexed. He took the discus-thrower of Myron as an example of a dynamic posture and pointed out that the arm holding the discus, the jaw limb, was stretched maximally. In seeking similar examples in the works of ancient Japanese artists, I was also able to find many examples in which the tonic neck reflex is expressed. Figure 1.35 shows a Buddhist statue, the Deva king (*Niō*) created by Unkei, which clearly shows extension of the jaw limb and flexion of the skull limb, thus giving a strong sensation of tension and energy. Figure 1.36 shows the God of Thunder drawn by a very famous old-time Japanese painter, Tawaraya Sōtatsu. Here, a perfect expression of the tonic neck reflex can be seen, with extension of the two jaw limbs and flexion of the two skull limbs. It is noteworthy that in this picture, not only is the head turned to the left but the eyes also deviate to the left. Furthermore, it is surprising to find that in the leg that is stretched, the jaw limb, the first toe is dorsiflexed and the other four toes are widely spread, coinciding with the Babinski position. In addition, all of the toes of the right leg which is bent, the skull limb, show plantar flexion. Thus the states of extension or flexion of every joint in the two legs are depicted in such a manner as to make the distinction between the jaw and skull limbs. Sōtatsu may have been aware of a possible physiological relationship between the tonic neck reflex and the Babinski position. Figure 1.37 shows a famous picture of a cat drawn by the late Japanese artist Takeuchi Seihō. The living sensation and lovely softness that may be felt in looking at the picture are indeed caused by the artist's keen observation of the cat and its expression (he is said to have observed the same cat every day for three months in order to draw this picture). He perceived the rule of the tonic neck reflex and depicted maximal extension of the cat's right forelimb in association with rotation of the head to the right.

I do not intend to maintain that all the famous sculptures and pictures

FIG. 1.35.   *Niō*, a guadian god (Kōfuku-ji, Nara).

of human and animal postures have been created in accordance with the posture of the tonic neck reflex, or that these works are artistically excellent because they coincide with this rule. I merely would like to emphasize that when one looks at a posture that is consistent with the tonic neck reflex, even one such as the mannequin shown in Figure 1.18, one experiences a natural, living sensation, because this posture coincides with the basic pattern of positional relationships between the head, trunk, and four limbs. I do not mean to insist that human movements must conform to the tonic neck and labyrinthine reflexes, because humans can voluntarily move any part of the body to produce any combination of positions within the limits of the joints. There are in fact many postures in which

FIG. 1.36.   *Fūjin-Raijin-zu* (Kennin-ji, Kyoto).

FIG. 1.37

the jaw limb is flexed and the skull limb extended, such as some dancing postures and those in various pieces of sculpture.

The point I wish to make is that the tonic neck reflex, tonic labyrinthine reflex, and righting reflex are acting in normal healthy adults as basic patterns which regulate postures and movements, and that whenever humans desire to exert their maximal muscular strength, as in catching a ball (Fig. 1.6) or in a game of soccer (Fig. 1.9), they can do so by performing voluntary movements in accordance with these basic patterns.

Athletes also adopt various other postures that are not consistent with these basic patterns. These postures seem to be taken on one of the following occasions: (1) when the person who is assuming a posture in accordance with the tonic neck reflex suddenly makes a quick motion, as in fencing (Figs. 1.28 and 1.29), catching a ball (Fig. 1.32), or returning a ball in tennis (Figs. 1.30 and 1.31); (2) when a person has to assume a posture other than that of the tonic neck reflex because of the nature of the action he wants to perform, as in a backhand stroke in tennis (Fig. 1.33); (3) when a hitherto unknown reflex is elicited. When Magnus undertook his original research, the famous work of Sherrington (1920) on reciprocal innervation had already been published, and later, still other new reflexes were discovered by de Kleijn and Versleegh (1930). Thus I believe that there are still many unknown reflexes to be discovered.

1.7.2. Interrelation of Tonic Neck, Tonic Labyrinthine, and Righting Reflexes

The relation of the tonic neck reflex, tonic labyrinthine reflex, and righting reflex in normal healthy adults has next to be considered together with the role of the labyrinthine organ in the formation of postures in man and animals in general. Humans and animals all have a specific normal head position and a normal body position relative to the horizontal. Man assumes these normal positions of the head and body when he is erect. If a passive movement is added to this standard posture (e.g., if the balance of his body begins to be lost by tilting of the plane on which he is standing, or if he stumbles while running, or loses balance while walking on a horizontal bar, as in Figure 1.24), his eyes and head and then his body rapidly return to the normal positions; that is, the righting reflex acts instantaneously. In rabbits, guinea pigs, and some other animals, the righting reflex is elicited by impulses of labyrinthine origin only, whereas in dogs, cats, and higher mammals this reflex can be evoked even if both labyrinths are destroyed, while leaving the visual organs intact. In man this reflex is evoked by both labyrinthine and visual impulses. In the daily postures of normal adults, the righting reflex functions in the maintenance of equilibrium. However, if a man performs active movements, such as turning, diving, or standing on his hands, he does so by voluntarily moving his head and body out of their normal positions. During these movements his head position changes markedly relative to the horizontal and to his trunk. Two factors have to be considered in analyzing the head position in such postures: the head position relative to the horizontal, and its position relative to the trunk. For any given position of the head, the musculature of the trunk and the four limbs is affected by both tonic neck and labyrinthine reflexes. In experiments on decerebrated cats, Magnus demonstrated that the combined effects of the tonic neck and tonic labyrinthine reflexes could be expressed in terms of an algebraic total. He found that when the head was in such a position that both reflexes acted to increase the extensor tone of a limb, the limb was extended markedly, whereas, when the head was in such a position that both reflexes had an attenuating effect on the limb's extension, the attenuation of the extensor tone was increased. He also demonstrated experimentally that if the position of the head was such that the two reflexes had opposite influences on a limb, there was no detectable change in the muscle tone of the limb, because the two opposing influences cancelled each other out. He reported that the same phenomena could be observed in normal animals. This problem will be discussed below with examples of human postures.

As already described, the extensor tone of the four limbs is maximal in man and the cat (cf. Fig. 1.4) when the angle between the oral fissure and the horizontal is $+45°$, and is minimal when the angle is $-135°$. If this reflex is considered to be the prototype of the human tonic labyrinthine reflex, and if this same reflex is assumed to exist in normal healthy adults, the extensor tone must increase when the head of a standing man is dorsiflexed from the normal position and must decrease when the head is ventroflexed.

Let us consider this problem with reference to figures. In the postures of gymnastic exercises in Figures 1.21 and 1.22, the dorsiflexed head position is accompanied by extension of the four limbs, and the ventroflexed head position is accompanied by their flexion. In these cases, the positions of the head, trunk, and four limbs are in accordance with both the tonic neck reflex and the tonic labyrinthine reflex, and these positions can be thought of as the algebraic sum of the two reflexes.

Next with regard to the posture of the man standing on his hands (Fig. 1.23): the righting reflex is activated by impulses of labyrinthine and visual origin so that the normal head position can be recovered as nearly as possible. Consequently, the head is markedly dorsiflexed relative to the trunk, and by means of the tonic neck reflex thus provoked, the four limbs become extended and the arms become capable of supporting the weight of the body. On the other hand, when the head position is analyzed in terms of the tonic labyrinthine reflex, the angle between the oral fissure and the horizontal is nearly $-135°$, and the position of the head relative to the horizontal is that in which the extensor tone is minimal. Thus in this case, just as in the preceding example, the stretching effect of the tonic neck reflex should be counteracted by the flexing effect of the tonic labyrinthine reflex, if Magnus's rule of cancellation is true. It is not possible to stand on the hands if the head is not dorsiflexed.

Thus I believe that the fundamental principle of standing on the hands consists of two separate, though related, movements. First, the head is raised by impulses of labyrinthine and visual origins, and then, aided by the tonic neck reflex thus provoked, the arms stretch maximally. In this case, contrary to normal, the tonic labyrinthine reflex must be inhibited. If, when standing on the hands, the head were not raised but remained in the same plane as the trunk—in other words, if the standing posture of the body were turned through 180° around the bitemporal axis—the extensor tone would increase, due to the effect of the tonic labyrinthine reflex, assuming that this was acting, since the angle between the oral fissure and the horizontal would become about $+45°$, but the tonic neck reflex would not be induced, since the head would remain in the same plane as the trunk. To maintain the head in the same plane as the trunk while standing on the hands requires training and skill and is difficult, if not impossible, for an ordinary person. If, while standing on the hands, the head was ventroflexed so as to assume a prone position in relation to the trunk, the angle between the oral fissure and the horizon would approach more closely to $+45°$, and the extensor tone would tend to be increased still further by the tonic labyrinthine reflex, but decreased by the tonic neck reflex. As a matter of fact, to the best of my knowledge, not even a very skillful acrobat could stand on his hands with his head in such a position.

From examination of the interrelationship between the tonic neck reflex and the tonic labyrinthine reflex in human postures, it can be concluded that the effects of the two reflexes on the muscular tone show an algebraic summation only in cases where their effects augment each other in increasing muscular tone: in cases where their effects oppose each other, this algebraic summation cannot be observed. Further, generally speaking, in the latter cases, the effect of the tonic neck reflex plays the principal role, while that of the tonic labyrinthine reflex is very weak.

In summary, the tonic neck reflex is certainly an important basis for momentary dynamic postures. The righting reflex, labyrinthine reflex, and visual reflex no doubt also play an important role in maintaining human postures, but the tonic labyrinthine reflex does not seem to be very important. I do not mean to say that the tonic labyrinthine reflex is an unnecessary postural reflex, or that the vestibular labyrinth is not important in human movements or human postures: simply that little is yet known about the role of the labyrinth in reflex regulation of human dynamic postures.

1.7.3. Two Phases of the Labyrinthine Reflex

I believe that otologists have hitherto paid excessive attention to experimental eye nystagmus evoked by caloric or rotatory stimulation and have rarely studied the vestibular labyrinth in terms of postural reflexes. As I have pointed out in a previous paper (Fukuda, 1959b), the equilibrium of the body is disrupted during the period of manifest, experimental nystagmus that is accompanied by such signs as severe rotatory vertigo, a marked Romberg's phenomenon, and the inability to stand or walk. In other words, this condition can be called transient artificial Ménière's disease, caused by excessively strong labyrinthine stimulation.

It may seem strange to discover that the vestibular labyrinth, an organ which has the specific role of maintaining body equilibrium, gives rise to experimental nystagmus and ultimately disrupts body equilibrium when stimulated excessively. When a patient shows spontaneous nystagmus accompanied by vertigo, it is usually thought that these signs indicate impairment of the inner ear due to otitis interna or Ménière's disease. When such signs of body imbalance as nystagmus or vertigo are induced experimentally by rotatory or caloric stimulation, the labyrinthine reflex is thought to be physiologically normal, and the subject's labyrinth is concluded to have normal equilibrating ability. How illogically we have been thinking!

However, if a rotatory stimulus that is too weak to induce experimental nystagmus is applied, or the pressure in the external auditory meatus is changed to stimulate the labyrinth slightly (these stimuli may not be weak, but must be most physiological for the labyrinth), the vestibulospinal reflex is provoked to maintain the body equilibrium. I proved this by the blindfolded writing test and the stepping test and named it the stage of coordination of the labyrinthine reflex, as contrasted with the stage of disturbance, which is characterized by unstable equilibrium and the appearance of experimental nystagmus (Fukuda, 1959b). The labyrinthine reflexes hitherto known as *Gegenrollung* (counter rotation) of the eyeballs, the righting reflex of a blindfolded man provoked by von Stein's goniometer, or a tilting table, and *Progressivbewegung* (progressive movement) or *Sprungbereitschaft*, which is observed when the body moves linearly, are labyrinthine postural reflexes in the stage of coordination.

As I stressed above, it is necessary to recognize the presence of two entirely different phases of the labyrinthine reflex: the stage of coordination and that of disturbance. In reviewing research on labyrinthine physiology, one finds that since the time of Bárány, physiologists have placed undue emphasis on experimental nystagmus, a pathological sign evoked in normal persons, and the stage of disturbance; the stage of coordination, or the proper vestibular labyrinthine function, has long been neglected. I believe that there have been no investigations of the latter phase since those of Magnus.

I would like to emphasize again the necessity for extensive and thorough study of the stage of coordination of the labyrinthine reflex. Furthermore, I postulate that the vestibular labyrinth, in association with the visual organ, plays a leading role in the composition of human postures. Here I would like to explain my reason for this postulation with some examples.

When a blindfolded cat or rabbit is held upside down in the air and then dropped, it instantly turns its body over and lands on its feet. Cinematographic examination of these rapid reflex movements of the falling animal shows that they are initiated and completed by the vestibular labyrinth. Initially, the head turns to its normal position relative to the horizontal as a result of the labyrinthine righting reflex. For this, the head is rotated relative to the trunk, and consequently, the tonic neck reflex occurs, i.e., the jaw forelimb is stretched while the skull one is bent. As a result of these positions of the forelimbs, the trunk easily turns over to regain its normal position due to the effect of the neck righting reflex on the trunk (Magnus, 1924). Next, the forelimbs and hind limbs assume symmetrical, lightly flexed positions due to the labyrinthine reflex *Progressivbewegung* during the fall through the air, and finally the animal reaches the ground with its head, trunk and four limbs in the normal positions. Thus the vestibular labyrinth initiates and regulates these reflex movements by evoking the tonic neck reflex, etc., in the course of these movements. In other words, these reflex movements depend entirely upon the integrity of the labyrinth.

This theory that the vestibular labyrinth is of great importance in human postures is supported by findings in an experiment on blindfolded leghorns (Fukuda, 1959a). In this experiment, birds that had received repeated rotation training showed different head positions during rotation from those they showed before training, thereby indicating development of vestibular function.

1.7.4. Howorth's Basic Dynamic Posture

Before ending this section, reference should be made to an interesting paper on human dynamic postures by Howorth (1946). Using cinematography, he examined and analyzed the various movements of babies and infants and various postures observable in adults in normal daily movements, and in a variety of sports. He postulated the existence of a basic dynamic posture, or basic dynamic position (Fig. 1.38). He stated, "The basic dynamic posture is characterized by a slight crouch, with the ankles,

FIG. 1.38.   Basic dynamic posture.

knees and hips flexed, the head and trunk inclined forward and the trunk
slightly flexed, the arms relaxed and slightly flexed. With the body in
this position, the muscles are in a mid position with an increased tone,
balanced, and ready for instant and powerful action in any direction."
He considered that this posture is basic to all types of motion. When
viewed in the light of the postural reflexes described above, his conclusion
seems to be quite correct. The reason why Howorth named this posture
the basic dynamic posture is that it seems to underlie all the various postures he observed. This posture is also basic from the standpoint of
postural reflexes. Goldstein (1923) states in connection with his *induzierte
Tonusveränderung* (induced change in tonus) that the optimal condition
for eliciting a reflex is to have the musculature of the entire body in a state
of slight tension and a considerable degree of mobility; namely, in a state
in which the postural reflexes can be most efficiently and perfectly induced.
Various types of postural reflexes, such as the tonic neck reflex, tonic
labyrinthine reflex, and righting reflex, can best be elicited by first assuming Howorth's basic dynamic posture before following the patterns
of the various reflex types. In this way the maximal muscular force can
be produced. Howorth's basic dynamic posture, or position, is also the
basis for efficient dynamic postures when considered from the standpoint
of postural reflexes.

Human movements are activated and postures are assumed initially by
way of the pyramidal tract, that is, by volition. However, most subsequent

processes are performed by way of the extrapyramidal system. The vestibular labyrinth, together with the visual organs, plays a leading role in this process by purposefully controlling and modifying many proprioceptive and exteroceptive reflexes. As I have shown above (when discussing the tonic neck reflex, tonic labyrinthine reflex, and righting reflex on the basis of this postulation of mine), though most human dynamic postures are initiated voluntarily, they are composed of various reflexes.

Human movements differ from those of animals in that the waist plays an important role. It is therefore necessary for us to consider the lumbar reflex, which is activated by changes in the relative positions of the waist and legs. This problem, together with other important problems of human dynamic postures, such as the physiology of training, demonstration of the unique role played by the vestibular labyrinth during the process of adaptation to certain special environments, and the relationship between the Babinski posture and the tonic neck reflex (as seen in the picture by Sōtatsu, Fig. 1.36), will be discussed in Chapter 2.

Chapter 2

The Babinski Reflex and Posture

2.1. Pathology and Physiology of the Babinski Reflex

About 80 years ago, Babinski discovered that when the sole of the foot is stroked, the big toe is dorsiflexed if the pyramidal tract is impaired. The orthodox Babinski reflex is simply dorsiflexion of the big toe, but spreading out (fanning) of the remaining toes is also often observed. In the original test, the edge of the sole of the foot was stroked (Fig. 2.1), but

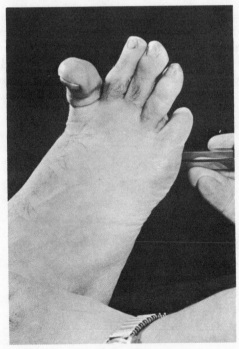

FIG. 2.1. Eliciting the Babinski reflex. Reproduced from Spillane (1975).

41

the reflex can also be elicited by pressing the tibia (as per Oppenheim) or by pressing the skin below the internal malleolus (as per Chadock). It should be noted also that when pyramidal disturbance is marked, the reflex is elicited by stimulating other parts of the body: by pricking the cheeks, for example. It is noteworthy that various workers have shown that the reflex can occasionally be observed in normal adults who do not have any pyramidal disturbance. Furthermore, Tokizane *et al.* (1951) found by EMG studies that when the sole of the foot of a normal individual was stroked, the tonus of dorsiflexion was increased; that is, the Babinski reflex was elicited latently. In other words, they showed that the Babinski reflex was not a pathological reflex elicited only in cases of pyramidal disturbance, but that it was a normal physiological reflex.

What significance does this physiological Babinski reflex have? I will evaluate this question from the standpoint of human postures (Magnus's *Körperstellung*).

2.2. Physiology of Rowing

The Babinski reflex is also clearly elicited while rowing a boat. The rower first holds an oar with the upper limbs extended, then flexes them maximally, stretches the lower limbs and pushes them out maximally, and during this last movement the Babinski reflex is observed in the feet on the stretch board (Fig. 2.2). The more vigorous the movement of the arms and legs, the more clearly is the reflex evoked. The details of the

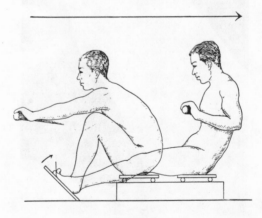

FIG. 2.2

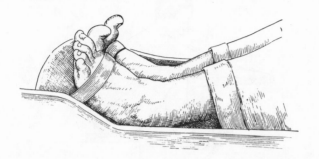

FIG. 2.3

movement of the legs are as follows: the legs are first fully flexed at the knee joints and then the knee joints are extended maximally up to 180°, so as to push out the legs. During extension of the legs, the Babinski reflex is elicited and the big toe is dorsiflexed maximally (Fig. 2.3). In this situation, the Babinski reflex has no significance as a defense reflex: it is a postural reflex elicited when the lower limbs are about to be extended fully.

The Stretch Board for Rowing: Mr. Kimisaburo Takagi is a friend of mine who was the coach of the boat club of Kyoto University. Through intuition and long years of experience, he managed to improve the stretch board for rowing, thereby increasing the speed of his crews tremendously.

When he first started coaching, the stretch board was simply a flat piece of board for the five toes. His improvement was to add a wedge-shaped piece of wood to the flat board, as shown in Figure 2.4, so that the toes, and particularly the big toe, could be dorsiflexed at 30–45°. In other words, he modified the stretch board so as to allow dorsiflexion of the toes, which induced the *supporting reflex* (Magnus's *Stützreaktion*) and reinforced the extension of the lower limbs. This modification increased the force of lower limb extension, the rowing power, and the rowing time. This is a good example of how important this postural reflex is in exertion of muscular force.

Those who have ever rowed a boat will probably remember feeling increasingly unsure about their toes when they pushed out their legs hard immediately after pulling the oars back. This is the time when the Babinski reflex is elicited (Fig. 2.2). When the legs are fully pushed out

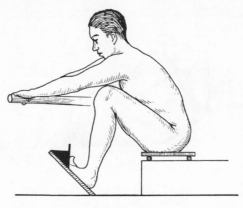

FIG. 2.4

for rowing, the toes are not really touching the stretch board. This position of the toes is assumed so as to produce lower limb extension by means of the Babinski reflex and the supporting reflex. However, the force exerted by pushing out the toes is lost because the toes do not touch the board. Mr. Takagi's improvement is reasonable in the sense that the wedge added to the board results in increase in the force during extension of the legs by making full use of the force of the toes when they are pushed out. Thus it increases the efficiency of rowing.

2.3. Physiology of High Heels

Mr. Takagi told me that his idea for improvement of the stretch board had been inspired by ladies' high-heeled shoes. The foot in a high-heeled shoe is dorsiflexed at the toes (Fig. 2.5). As the X-ray picture shows, the toes are dorsiflexed at the second phalanges (Fig. 2.6). The toes are forcibly arranged by the high heel to be dorsiflexed so that the supporting reflex is evoked, which in turn increases the lower limb extensor tonus. The value of high heels is understood well in terms of the Babinski reflex and the supporting reflex; these shoes are not useful just to make a person look taller. They increase the lower limb extensor tonus, thereby allowing a straight, upright posture. High heels were invented on the basis of experience, but have a physiological significance.

When strong extension of the legs is required, the heels are raised off the ground and the toes are fully dorsiflexed. Examples of this can be seen in the feet of a runner or of a person pulling a boat against a current.

FIG. 2.5

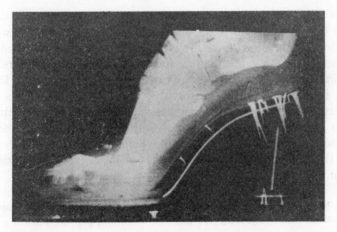

FIG. 2.6

This position increases the pushing strength of the legs by dorsiflexing the toes maximally. The straw shoes of the boat men on the River Nagara who pull pleasure boats against the strong current are simple footgear composed of straw mats on which only the toes can be placed.

2.4. Reflex in the *God of Thunder* by Sōtatsu

Mention has already been made of the picture of the God of Thunder by Sōtatsu (Fig. 1.36). As I have stated in Chapter 1 concerning the statue of the Deva king created by Unkei in connection with the tonic neck reflex, the head of the God of Thunder is also turned. The upper and lower limbs on the side of the chin are fully extended, whereas the two limbs on the occipital side are markedly flexed; that is, it is the tonic neck reflex itself that gives the impression of power. Here I would like to point out the position of the eyes, which I did not mention earlier. The eyes of the Thunder God are turned down and to the left. Like the head position, the eye position changes the muscular tonus of the limbs characteristically. Turning the eyes to the left, as seen in this picture, evokes the same reflex of the limbs as turning the head to the left, thereby increasing extension of the left limbs and flexion of the right ones. In other words, the tonic neck and eye reflexes give rise to marked extension and flexion of the lower limbs, as seen in the picture of the God of Thunder.

Now let us move on to the plantar reflex. Attention should be directed to the toes of the lower limbs of the God of Thunder, one of which is extended and the other flexed. The left leg, which is fully stretched, has a dorsiflexed big toe with the other four toes fanned out and dorsiflexed. Drawing the five toes in this way expresses the power of the left leg, which is extended and thrust forward. Conversely, all five toes in the right leg show plantar flexion, emphasizing the flexion of this limb.

It is remarkable that Sōtatsu not only perceived the tonic neck reflex and the eye reflex, which affect the limbs, but also apparently understood the physiology of the plantar reflex of the big toe, dorsiflexion and plantar flexion.

Note: Much of the history and literature on the Babinski reflex in this chapter was taken from Wartenberg (1947).

Chapter 3

Postures and Vertigo

3.1. Introduction

Seasickness has been known ever since man began to sail the seven seas, but with the coming of airplanes, the sickness experienced in various vehicles has been considered under the wider concept of motion sickness, and its prevention and treatment have become important subjects of study.

A point that must be taken into careful consideration in research on this problem is that when motion sickness or a similar condition is evoked for observation, the examinee or animal is always moved passively. For instance, a man is fastened into a rotatory table and turned round or shaken up and down, and in both treatments he is passive. What seems strange is that even during fairly vigorous motion in non-experimental conditions, motion sickness is never produced if the person or animal is moving actively. This fact should receive more attention. Nobody has heard of a rabbit developing motion sickness when it is running around in the mountains and fields, although it receives various labyrinthine stimuli from the linear or circular motions and jumping it performs. When a child skips rope, the labyrinth receives considerable vertical stimulation. It receives additional linear stimulation if the child runs while skipping. In spite of these stimuli, the child does not suffer from motion sickness. In other words, motion sickness is a condition brought on a person or animal who is moved passively. I think that it is essential to classify motions into two classes, active and passive. In ordinary experiments on labyrinthine physiology, animals are fixed in position and moved vertically or rotated for observation of labyrinthine reflexes in the eye muscles or skeletal muscles. Physical stimuli such as vertical movement or rotation should affect the otolith and cupula in the same way, whether they are received passively, as in a laboratory experiment, or actively, as in normal human or animal movements. The fact that the same stimulus evokes vertigo far more easily

47

FIG. 3.1

when given passively suggests that the vertigo cannot be explained in terms of the physical effect of labyrinthine stimulation and the resultant reflex. I believe that this is probably due to a difference in muscular tone between passive and active motions. In the following sections I will discuss the problem of motion sickness and human postures, as well as that of postures taken while controlling vehicles and the effects of motion, in terms of active and passive motions and movements.

Before beginning on the main theme, let us review an interesting paper that deals with prevention of air sickness. This paper, entitled *Body Position and Motion Sickness*, was written by C. M. Manning and W. G. Stewart of the Aviation Medicine Research Institute of Canada, and the work described was prompted by the fact that paratroopers frequently developed air sickness and thus became unable to act efficiently on landing.

In this study, soldiers were placed on a large swing and asked to assume various postures (Fig. 3.1) so that the stimuli received by the labyrinth could be analyzed in detail. Manning and Stewart concluded from their laborious studies that to avoid motion sickness, the soldiers should be transported by air in the supine position with their eyes open. In this position vertigo is minimal since the vertical semicircular canal becomes horizontal. I found it interesting that other people besides myself were studying the subject of motion sickness and posture. Manning and Stewart studied the posture in which motion sickness was minimal; I myself will discuss this subject of postures and motion sickness in terms of active and passive motions, which produce very different effects: passive motion results in vertigo, whereas active motion does not.

3.2. Mirror Image Postures in Active and Passive Movements

Güttich (1940) noticed a very interesting thing when observing active and passive rotations. He found that the movements of the eyes and the head produced in active rotation were mirror images of those produced in passive rotation, such as when the examinee was rotated in a chair. As shown in Figure 3.2, during active rotation to the left, the eyes turn to the left, i.e., in the direction of rotation, and this is followed by movement of the head and trunk. However, in passive rotation, the eyes first move in the direction opposite to the direction of rotation (slow phase of nystagmus) and then quickly move back to the front as rotation

FIG. 3.2.   Active (left) and passive (right) rotations.

FIG. 3.3

FIG. 3.4

continues. In other words, in active motion, the eyes move in the direction of rotation and other parts of the body follow, whereas in passive motion, the eyes move in the direction opposite to the direction of rotation. The movements of the eyes in the two cases are mirror images of each other. That is, the movements are symmetrically opposite. The movement of the head relative to the trunk is also symmetrically different in the two cases, as shown in Figure 3.2. Thus Güttich saw that the eyes, head, and trunk move in symmetrically opposite directions in active rotation and in passive rotation.

I found that this was true not only in rotation but also in linear motion. When comparing the postures of the rickshaw-man and the passenger,

FIG. 3.5

one finds that the actively running rickshaw-man leans foreward, i.e., in the direction of running, whereas the passenger, who is being carried passively, leans beckward, i.e., in the opposite direction to motion (Fig. 3.3). Figure 3.4 shows a ski jumper with his body leaning forward like the rickshaw-man, while Figure 3.5 shows a water-skier with her body tilted backward in passive movement. These two postures form symmetric mirror images of each other. Thus the basic forms in active and passive linear motions are symmetrically opposite, just as Güttich recognized a symmetrical difference between active and passive rotations.

Attention should be paid to the fact that active and passive movements produce such mirror-image postures, as well as to the fact that passive stimulation evokes vertigo, whereas active stimulation does not, although the two stimulations seem to have the same effect on the labyrinth.

3.3. Posing Two Questions—Motion Sickness and Training

As mentioned above, active and passive motions produce different results with respect to vertigo and posture. These two phenomena seem to be related in many ways, but there are various unresolved questions with regard to their relation. In thinking about this matter, I have formulated two main problems, and I will present my own interpretation of this subject by discussing these problems.

The first problem is concerned with motion sickness. Motion sickness is experienced by passengers, be it on buses, trains, or boats. One never hears of a bus driver, a helmsman, or a train driver getting motion sickness. Soldiers transported by air may get airsick, but not the pilot. The people who are moved passively become sick, but those who are driving, steering, or navigating do not. Those who are in charge of the vehicle are certainly used to shaking, and psychological tension may also contribute to their being free from motion sickness. However, one often hears about bus drivers who are used to shaking and do not get sick while driving the bus, but unfailingly become sick when travelling as passengers on the same route. This cannot be explained in terms of getting used to the trip, and I would like to explain it without using an obscure idea such as psychological tension.

The second problem concerns the aptitude and training of paratroopers and other soldiers in airplanes, which we have discussed before. Findings on this subject have been applied on a larger scale by various organizations with great success. These findings are, first, that athletes who are trained in various active motions have the aptitude to become good pilots, and second, that people can be trained not to develop motion sickness in airplanes, which produce passive shaking, rotation, and vertical motions, by exercises such as self-rotation and jumping up and down. Aside from learning the immediately-necessary techniques of steering and control of the aircraft, it is not necessary to ride in airplanes for long periods to become accustomed to them and overcome air sickness: active exercises on the ground, such as various rotatory motions, have been found to be the best training.

In short, my two questions are: (1) Why do passengers develop motion sickness, while drivers do not, when they are shaken in the same way? and (2) Why are active exercises effective training for overcoming the motion sickness produced by passive movement in airplanes? These questions stem from the observations that active and passive movements are different in that the postures resulting from them are symmetrically opposite, and that passive movements induce motion sickness, whereas active ones do not. I will attempt to answer these questions in later sections.

3.4. Postrotatory Nystagmus Evoked by Changes in the Position of the Eyes and the Head

The following two tests were made on 100 healthy men and 100 healthy women who were free from otological diseases. The examinees were

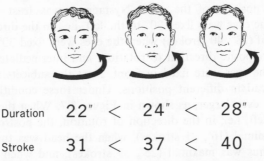

Duration 22" < 24" < 28"

Stroke 31 < 37 < 40

FIG. 3.6. Duration and stroke of postrotatory nystagmus after passive rotation with eyes fixed to the left (left), in front (center), and to the right (right). Mean values of 100 healthy subjects.

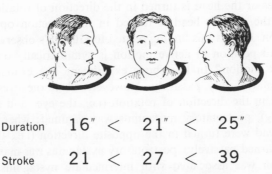

Duration 16" < 21" < 25"

Stroke 21 < 27 < 39

FIG. 3.7. Duration and stroke of postrotatory nystagmus after passive rotation with head turned to the left (left), in front (center), and to the right (right). Mean values of 100 healthy subjects.

placed on a revolving chair and turned around passively. Following Bárány's method (Bárány, 1907), each subject was rotated ten times in 20 seconds, and the duration and frequency of postrotatory nystagmus were timed and counted. The experiment was repeated three times with the eyes and head in different positions, and the postrotatory nystagmus was recorded each time. First the examinee was asked to look at an otogoniometer, which is a small mirror hung in front of him, to fix the position of his eyes during rotation. In this way, the position of the eyes was fixed, at 30° to the right or 30° to the left during the ten rotations in 20 seconds. Then the postrotatory nystagmus was recorded. As shown in Figure 3.6, the duration and frequency of nystagmus varied depending on the position of the eyes. Nystagmus was least (22″, 31 strokes) when the eyes were fixed 30° to the left, i.e., in the direction of rotation, maximal (28″, 40 strokes) when the eyes were fixed 30° to the right, i.e., in the opposite direction to rotation, and intermediate (24″, 37 strokes) when the eyes were fixed in front. Next the subjects were rotated with their head in different positions. Under these conditions the variation became even larger, as shown in Figure 3.7. When the head was turned to the left, i.e., in the direction of rotation, the postrotatory nystagmus was minimal (16″, 21 strokes), when the head was turned to the right, nystagmus was maximal (25″, 39 strokes), and when the head was not turned, nystagmus was intermediate (21″, 27 strokes). In other words, postrotatory nystagmus differed markedly depending upon the position of the eyes or the head. What do these data mean?

The data shown in Figures 3.6 and 3.7 can be summarized as follows: the duration and frequency of postrotatory nystagmus are reduced when the eyes or the head is turned in the direction of rotation, and increased when the eyes or head is turned in the direction opposite to that of rotation. How is this finding related to Güttich's observation that the eye or head position in active motion is symmetrically opposite to that in passive motion? In the above experiment, the examinee on the chair was rotated only passively. However, when the eyes and head were turned in the direction of rotation (i.e., the eyes and head were turned actively), postrotatory nystagmus was minimal; whereas when the eyes and head were turned in the opposite direction (i.e., the eyes and head were turned passively), postrotatory nystagmus was maximal. The frontal position was associated with intermediate nystagmus. During postrotatory nystagmus, vertigo and dysequilibrium persist for about the same time as nystagmus. In other words, although the rotation is passive, if the examinee assumes an active posture (i.e., with the eyes and head in the direction of rotation), postrotatory vertigo and dysequilibrium will probably be minimal, as postrotatory nystagmus persists for the shortest period. With these findings and ideas in mind, I will discuss my first question.

3.5. An Answer to the First Question

We know from experience that passengers get motion sickness whereas drivers do not, and that a person who does not get car-sick when driving himself gets sick in a car driven by somebody else. These people are in the same vehicle where they are apparently being moved passively in the

FIG. 3.8

same way and yet their reactions are different. Let us consider this problem by examining people's postures in a bus. Figure 3.8 shows the different postures people assume when a bus turns to the right. The driver is turning the steering wheel to the right and is tilted to the right with his head and trunk assuming a centripetal posture relative to the circular movement of the rightward turn. When one makes a turn while actively running, one's head and trunk are tilted towards the center of the curve as the body resists the centrifugal force. The bus driver, although being carried by the bus, that is, moving passively, is assuming the same posture as he would were he running actively. In contrast, the passenger standing on the right has his head and trunk tilted in the direction opposite to that of the driver by the centrifugal force of turning. His posture is symmetrically opposite to that of the driver. These mirror-image postures are found not only between a bus driver and a passenger but also between a train driver, a helmsman, or an airplane pilot and their passengers: the former assume active postures, whereas the latter assume passive ones. Passengers are at the mercy of vertical, rotatory, or linear motions and particularly changes in these motions; that is, they assume passive postures under these conditions. Drivers, on the other hand, move their bodies actively while they themselves are being carried by the vehicles they are driving. These two groups of people have postures that are symmetrically opposite, i.e., mirror images of each other. According to the data shown in Figures

3.6 and 3.7, the postrotatory nystagmus is shortest and smallest when the eyes and head are turned in the direction of rotation, that is, when they assume an active position. This is comparable to the posture assumed by the bus driver in Figure 3.8. Conversely, when the eyes and head are turned in the direction opposite to that of rotation, the postrotatory nystagmus is longest and greatest; this is comparable to the case of the passenger in Figure 3.8. Vertigo and dysequilibrium, which means transient motion sickness, are present when there is postrotatory nystagmus. Thus the fact that the driver develops little or no motion sickness can be explained by the fact that he assumes an actively moving posture in spite of the passive motion he is subjected to by the vehicle: that is, the driver assumes a posture that insures minimum postrotatory nystagmus. Passengers get sick because they are forced to take passive postures as the motion changes; i.e., they assume postures in which sickness is easily produced. This is my answer to Question 1.

People tend to rush to secure seats for themselves when getting on buses or trains. Japanese boats provide pillows even during the day so that passengers can lie on their back and allow their bodies to move with the boat. Why are these good postures in terms of postrotatory nystagmus? As shown in Figures 3.6 and 3.7, when sitting with the head kept straight, postrotatory nystagmus is less pronounced than when the head assumes a passive rotation posture, i.e., when the eyes and head are turned in the direction opposite to rotation. That is, with this head position, there is less vertigo. In the second situation of lying on the back on board a ship, the person is at the mercy of the motion of the ship, which is comparable to the position with the eyes and head facing straight forward. When a passenger assumes some unstable posture, however, his body will move in a direction opposite to that of the vehicle, which is either rotating, moving up and down, or moving horizontally. This means that in passive rotation, the eyes, head, and trunk are turned in the opposite direction to rotation, which results in the longest postrotatory nystagmus and, therefore, vertigo. Returning to Figure 3.8, the person standing in the bus while it turns is tilted to the left by the centrifugal force, whereas the seated passenger leans against the back of the seat resisting the centrifugal force, and thus maintaining a normal posture. The seated passenger is comparable to the person whose head and eyes are positioned straight forward and thus his postrotatory nystagmus is less than that of the standing passenger. That is, a sitting posture that moves with the vehicle produces less motion sickness than a standing posture. This explains why people try to get seats or lie down on their backs on board a ship. Man has learned from experience to assume these as defensive postures.

The following conclusions can be deduced by comparing Figures 3.6 and 3.7 (eye or head position and postrotatory nystagmus) with Figure 3.8 (passengers on a bus):

a) Postrotatory nystagmus is minimal with respect to duration and frequency when the eyes and head are turned in the direction of rotation.

b) The extent of nystagmus is intermediate when the subject is facing straight forward.

c) Nystagmus is maximal when the head turns in the opposite direction to rotation.

Of the above, a) is comparable to the driver, b) to the seated passenger, and c) to the standing passenger. This explains why people in a bus in different postures show different degrees of motion sickness, even though they are all in the same moving bus.

3.6. Physiology of *Man and Horse Together* and an Answer to the Second Question

The second question was why exercise is a useful training for getting used to travelling in airplanes, which force man to move passively. This can easily be understood in terms of the posture assumed by the bus

FIG. 3.9

FIG. 3.10

driver described above. Before discussing this further, I want here to consider the posture taken while riding on a vehicle and its kinetic effect. This means the kinetic effect when assuming an active posture during passive movement. Figure 3.9 shows a man riding on a motorcycle and Figure 3.10 shows a man on the back of a jumping horse. Both people are assuming an active kinetic posture of leaning well forward during their passive motion, as if they were themselves running. This is in contrast with the posture assumed by the passenger in a rickshaw, which was the mirror image of that of the rickshaw-man. This posture of active motion is particularly clear in the curve of Figure 3.9, in which the axis of the body is tilted inward centripetally. The assumption of such an active posture during passive movement is important for mastering the movement and for obtaining complete kinetic effects. As everybody knows, mastering such movement is inherently difficult: when learning to ride a horse, for example, one's body tends to be tilted backward and the lower part of one's body tends to slide towards the tail; when learning to ride a bicycle, even after one has mastered the technique of riding straight, one easily falls while turning a corner because one tilts one's body in the direction opposite to the curve. In other words, the secret of mastering riding a horse or motorcycle, skating, or skiing, in which the whole or part of the body is subject to passive movement, is in assuming active postures during passive movements. The expression *man and horse together* means that a man on horseback should tilt his body fully forward while being carried by the horse so that he "runs" himself while on horseback.

Thus Question 2 has been answered. To reiterate, the question was why is active exercise helpful for getting used to passive movement in an airplane, since in active and passive movements, the movements of the eyes, head, and trunk are mirror images of each other. The answer is that when the plane moves round or up and down, the best kinetic effect is obtained, and motion sickness is avoided, if the body (and therefore the eyes, head, and trunk) moves together with the plane.

3.7. Discussion

In this chapter motion sickness was analyzed in the light of the difference between active and passive postures. Yet, not everything about motion sickness has been clarified: vision and getting used to movements are also important. Vision can alleviate motion sickness, although visual stimulation can worsen it, as I have described elsewhere (Fukuda, 1958). Experiments on animals have shown that the duration and the frequency of postrotatory nystagmus decrease when animals become accustomed to repeated rotation. My co-workers and I have also demonstrated this in man (Fukuda, 1958). We also showed that postrotatory nystagmus is far less frequent in good athletes and airplane pilots than in normal people and that the former rarely suffer from motion sickness. The extent of adaptation to motion can be shown objectively in terms of changes in postrotatory nystagmus.

In previous physiological studies on the labyrinth, which plays an important role in motion sickness, only passive rotation and linear movement were examined, and function was analyzed on the basis of the reflex movements caused by these movements. Thinking that this approach was inadequate, I introduced the concepts of active and passive movements into studies on labyrinthine function and equilibrium. People with impaired labyrinthine function are quite clumsy or incapable of active movements. When they are made to rotate actively, they are incapable of turning around on a fixed point, and instead rotate in gradually enlarging circles, i.e., in a spiral, on account of their own centrifugal force. Since this phenomenon cannot be explained in terms of traditional labyrinthine physiology I proposed the concept of active movement.

Outside Japan, a study was carried out in an effort to prevent motion sickness in soldiers transported by air (Manning and Stewart). In this study, 825 soldiers were swung for 30 minutes while assuming 14 different postures to see whether they became nauseated, vomited, became pale, experienced cold-sweat, or collapsed. This study showed that motion sickness was least when the vertical semicircular canal was maintained horizontally with the eyes open. This proves scientifically the value of sitting in a deck chair on board a ship: the chair is inclined so that its occupant assumes this head position. This interesting approach of studying postural as well as medical means of preventing motion sickness should lead to a true understanding of motion sickness. There is one report from abroad in a newspaper on training people on swings to prevent seasickness, which I think is a very interesting method. From the time when a person begins to set the swing in motion until he finally makes it move in a wide arc, he is exerting active movements, while during the period when the swing is moving fully, he is subjected to passive movements. Different stimuli are given to the labyrinth when the eyes are closed or opened and when the position of the head or the eyes are changed. What is more remarkable is that both active and passive circular movements can be performed on a swing by the volition of the swinger: based on my theory mentioned above, this training on a swing should be very effective.

Chapter 4

Adjustment and Composition of Posture

4.1. Review of Posture Research

The following is a brief review of research on posture and a summary of recent studies. In old physiology textbooks, posture was evaluated from the standpoint of physical laws, the human body was treated as a single object, and its center of gravity was frequently determined. For instance, in an extremely old study recorded in *motu animalium* written by Borelli in 1651, Borelli made a person lie on a board and determined the point of balance, thereby studying the center of gravity. Research of this type continued until the end of the nineteenth century. Braune and Fischer (1895) froze five cadavers, which they severed in order to obtain the centers of gravity of the head, trunk, and upper limbs, and inferred the center of gravity of the entire body from their results. On the basis of this study, they defined *Normalstellung*, in which the centers of gravity of the head, trunk, and lower limbs form a straight line, as shown in Figure 4.1. They considered this as the fundamental upright posture. This posture was adopted by the Japanese army and is known to us as the posture of "standing at attention." Du Bois Reymond, a learned physiologist of those days, proposed the concept of *Standfestigkeit* and maintained that a person was *standfestig* while standing if the center of gravity fell within the area formed by the soles of the feet. This approach of treating a human body as an object and studying posture in terms of physical laws disappeared altogether from physiology textbooks in about 1910. The reason for this seems to be that it is not possible to understand the true essence of human posture by assuming that a human body is a rigid object, because it is flexible and constantly moving to assume unfixed positions, which enable quick movements. In other words, one cannot study human posture with an idea such as that a dish is more stable than a glass. This is why this approach could not advance any further.

There were, however, very important studies among these classical

61

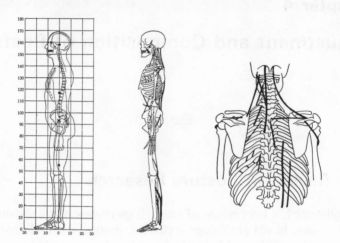

FIG. 4.1.   Left: *Normalstellung*. A dot indicates the center of gravity
of each part, and an x the center of gravity of the entire
body. Center and right: heavy black lines indicate the
antigravity muscles which function in standing.

physiological works. One of them was performed by Vierdordt (1881) and
Leitenstorfer (1897). They thought that the standing position was essentially quite unstable and that it could only be maintained by the presence of constant changes in skeletal muscular tension, i.e., reflexes. To
test this idea, they made a person wear a helmet, to the top of which
was attached a needle touching a smoked drum (Fig. 4.2). They thereby
demonstrated that in the standing position, i.e., the position of standing-
at-attention which had been considered to be still, there is not a moment
of stillness and that this position is maintained in constant fluctuation.
This is shown in their cephalogram (Fig. 4.3). In their study of the line of
the center of gravity, Basler (1929) made a person stand at attention on a
piece of board, one end of which was placed on a fulcrum. They reported
that the standing position was maintained not by absolute stillness but
by constant fluctuation as represented by continuous movement of the
line of the center of gravity (Fig. 4.4). In other words, they proved that
man has no absolute still-standing posture and that the standing position
is maintained by constant movement.

I would like now to analyze the structure of the nervous system which
controls human posture, assuming that the standing position is maintained by constant and delicate movement. Standing is a volitional movement controlled by the pyramidal tract. No matter how still one tries to

FIG. 4.2

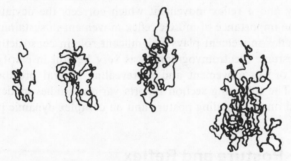

FIG. 4.3.   Cephalogram of a healthy subject standing with eyes open
for (left to right) 30 seconds, 1 minute, 2 minutes, and
5 minutes.

stand, one's position is constantly changing as described above. As a
human body is physically quite unstable, it tends to move according to
gravity, i.e., fall in such a way as to lower the center of gravity. This is the
physical cause of the initial fluctuation or deviation. In response to this
deviation, muscles quickly move the shifted center of gravity back to its

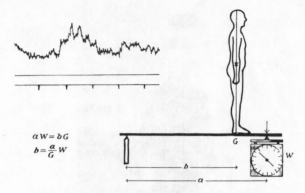

$$\alpha W = bG$$
$$b = \frac{a}{G} W$$

FIG. 4.4.   Reproduced from Basler (1929).

initial position. Although this movement is under the control of volition, the main role is played by subconscious muscle reflexes. The fluctuation seen in a cephalogram or that recorded by Basler can be considered to be composed of two factors: one is falling or deviation, tending to lower the center of gravity, and the other is reflex movement, which corrects the deviation in posture to regain the initial position. This fluctuation is a complex pendular movement around the line of the center of gravity, and is composed of a movement away from the line of the center of gravity and a reflex movement which corrects the deviation. Here one sees the importance of muscle reflex movement in sustaining the posture. This reflex movement plays a significant role in constructing and adjusting posture. Electromyography was very helpful in clarifying the mechanism of this movement and in revealing several previously unknown facts. The following section reports various studies made in Japan and abroad on the standing posture and on complex dynamic postures.

4.2. Posture and Reflex

As stated above, the standing posture is constantly maintained by muscle reflexes. How is this control achieved? The standing position is physically quite unstable and is subject to the pull of gravity. In other words, there is a constant physical movement tending to lower the center of gravity and move it away from the line of the center of gravity for the standing position, which results in disturbance in the standing position. What I call reflex movement or posture adjustment restores the center of gravity to the initial position by counteracting this downward movement. The receptors that sense the disturbance, disintegration, or deviation of posture and give rise to muscle reflexes have been identified in previous research: proprioceptors in the muscles and joints and exteroceptors in the skin give rise to muscle reflexes responding delicately to the destruction of posture: i.e., proprioceptive and exteroceptive reflexes take place. These proprioceptors and exteroceptors are distributed throughout the body. Their superior receptors are the labyrinth and the visual organs, which induce more important muscle reflexes and thereby participate in posture control.

Posture is constructed and controlled by muscle reflexes that are mediated by these receptors, the proprioceptors and exteroceptors, and the labyrinth and visual organs. Obviously, these reflexes are also controlled further by the central nervous system and especially the cerebellum. The reason for mentioning this is that I wish to point out that controlling or constructing postures and performing movements are whole-body movements and that they can only be implemented by integration of muscle reflexes induced by these receptors.

Of course, it is necessary and significant to study the mechanism of the reflex induced by each receptor. However, I feel that it is extremely important to pay attention to the cooperation of these receptors and the integration of the reflexes they induce rather than to just pay attention to individual receptors.

4.3. Standing Posture

The standing posture is an important fundamental posture of man. As already stated, classical physiological studies on posture placed emphasis on the posture of standing. Many studies were on the position of the center of gravity, and it was shown that the standing position was maintained in constant fluctuation. Electromyography contributed a great deal to studies on the standing posture by clarifying how skeletal muscles affect the standing posture. Before development of electromyography, it could only be conjectured which muscles or muscle groups functioned to counteract gravity. In those days it was only possible to palpate muscles and decide whether they were contracting from whether they were hard or soft. This was too crude a method. By electromyography it became possible to determine whether a muscle was contracting by inserting an electrode needle into the muscle and recording electromyographic spikes.

The muscles which function in standing, i.e., antigravity muscles, were found to be mainly those shown in Figure 4.1 and Table 1.1. It

TABLE 1.1. Antigravity Muscles and Intensity of Their Function

Part of body	Name of muscle	Intensity of function
Lower extremities and cingula	m. abductor hallucis	⧺
	m. abductor digiti quinti	±
	m. flexor hallucis longus	+
	m. flexor digitorum longus	+
	m. tibialis posterior	+
	m. soleus	⧻
	m. gastrocnemius caput tibiale	⧺
	caput fibulare	+
	m. biceps femoris caput longum	±
	caput breve	+
	m. gluteus maximus	+
	m. gluteus medius	+
Back, breast, and abdomen	m. sacrospinalis	⧻
	m. trapezius pars descendens	+
	pars transverse	+
	m. rhomboideus	±
	m. semispinalis capitis	⧺
	m. splenius capitis	+
	m. obliquus externus abdominis	+
	m. obliquus internus abdominis	+
	m. transversus abdominis	+
Upper extremities and cingula	m. deltoideus (pars spinata)	+
	m. supraspinatus	+
Head and neck	m. masseter	⧺
	m. temporalis	+

should be noted that they are mainly the back (or dorsal) muscles of the trunk and the dorsal muscles of the lower limbs. In other words, in the human body, the dorsal muscles are activated when assuming the standing position so as to correct the line of the center of gravity, which would otherwise tend to fall forward, thereby restoring the center of gravity to the center of the area of the soles of the feet (Hellebrandt, 1938; Braune, 1895). According to Hellebrandt (1938), the center of gravity in the standing posture lies 55% of the body height above the ground. It is interesting that the center of gravity is higher than one-half the body height. Thus the posture of standing is physically quite unstable and is barely maintained by the force or activity of the dorsal muscles. This physical instability facilitates motion and is the key to human and animal movements; complex movements are effected while maintaining a high center of gravity, utilizing the physical movement of falling according to gravity, i. e., a natural force, and exerting muscle forces in an antigravitational fashion. It can be appreciated from this standpoint why turtles,

FIG. 4.5.   The higher the center of gravity and thus the greater the
physical instability, the quicker the movement.

whose center of gravity is low and whose area of support is wide, are
slow and clumsy (Fig. 4.5).

As stated above, electromyography showed which skeletal muscles
function while standing. However, these muscles do not maintain a
constant, unchanging tension throughout the time of standing, thereby
holding the body rigid. The tension of the muscles varies and different
muscles function differently.

Some workers maintain that these antigravity muscles show little or
no electric activity when the body is standing. Floyd and Silver held
this view with respect to a back muscle, the sacrospinal muscle, and
Weddel, Feinstein, and Pattle, with respect to antigravity muscles of
the leg and thigh. Seyffarth went so far as to say that no action potential
could be recorded from such antigravity muscles of the leg as the gastrocnemius, soleus, and anterior tibial muscles. Hoeffer examined both
men and women. He reported that in three women the gastrocnemius
muscles did not move, while the anterior tibial muscles showed action
potential, and that in four men the gastrocnemius muscles moved, while
the anterior tibial muscles produced no action potential. These workers
concluded from these findings that the standing position is maintained
not by active muscular contraction but by the inherent elasticity of the
muscles.

However, from calculations based on the research of Hellebrandt
mentioned above, maintenance of the position requires a force of 16
kilograms by the dorsal muscles of the leg, i.e., the soleus and gastrocnemius muscles. Thus the standing position could not be maintained
solely by the elasticity of the muscles. Joseph and Nightingale (1952)
studied this issue from the standpoint of Hellebrandt and recorded action
potentials in these antigravity muscles. Their report is significant in that
they did not treat standing as a fixed concept but considered several
types of standing, such as standing at ease and standing with one leg
tensed. They also recorded electromyograms from various muscles. I

TABLE 1.2. Muscles Showing Spike Discharge in Joseph and
Nightingale's Experiment

Muscle	Normal standing	Standing mainly on the right leg	Standing mainly on the left leg
Right anterior tibial	0	0	1
Right soleus	12	12	1
Right gastrocnemius	7	6	0

have often found that when an experimental subject is ordered to stand, he tends to stand on the leg that does not have an electrode needle inserted, because the leg with the electrode is painful to use. So the leg with the electrode tends to be inactive and the muscles are over-relaxed and produce no recognizable spike discharges. The studies of Joseph and Nightingale are informative because these workers took various conditions into account. Their findings among 12 examinees of the numbers in which various muscles showed spike discharges are shown in Table 1.2. The same standing position produced different responses in different individuals, and different postures evoked different responses in different muscles. Their study showed that the standing position of man is maintained, not by muscle elasticity, but by the activity, or contraction, of antigravity muscles and especially the soleus muscle. It is noteworthy that during standing the gastrocnemius showed spike discharges in only half of the examinees, i.e., there was no fixed pattern. As will be shown later, when the arms are raised, different antigravity muscles of the lower limb become activated. Normally the ventral muscles of the lower limb are not activated while standing, but when the position of standing is shifted slightly forward or backward, the ventral muscles become activated. In other words, the antigravity muscles normally used in standing are the dorsal muscles mentioned above, but all skeletal muscles are capable of functioning in an antigravity fashion as the posture changes or as various movements are made. Man moves swiftly by utilizing the two forces, i.e., gravity and the antigravity action. I think that every muscle is capable of an antigravity action just like those of the antigravity muscles involved in the standing position.

The above considerations concern antigravity muscles of the leg, but Joseph and Nightingale reported that the thigh muscles also have an antigravity action. They described that although the biceps femoris muscle is generally regarded as the antigravity muscle of the thigh, it did not always function and that this muscle was active in only three of 14 examinees who were standing in a normal fashion. However, they recorded strong spikes from the biceps femoris muscle when the subject was standing with the upper limb raised forward or was leaning forward. In other words, maintenance of equilibrium depended on the standing position: when the center of gravity was shifted forward, a dorsal muscle of the thigh, the biceps femoris muscle, was strongly activated to maintain equilibrium. In short, posture is controlled by contraction of different kinds of antigravity muscles to different extents depending on the posture. This change in tension of antigravity muscles is a kind of reflex and can be termed an equilibrium reflex, postural reflex, or posture control reflex. The receptors which give rise to this reflex include the proprioceptors, the exteroceptors, the labyrinth, and the visual organs, as mentioned earlier. When the arm is raised forward, spikes appear in the biceps femoris muscle via the proprioceptors existing in the muscles and joints of the arm. In this way, the activity of the biceps femoris is a proprioceptive reflex. This corresponds to *induzierte Tonusveränderung* of Goldstein (1923). When the body is leaning forward, changes are induced in the proprioceptors of other muscles as well as those in the biceps femoris muscles themselves. In addition, the labyrinth and the visual organs are stimulated as the head position changes when the body leans forward. Conceivably they also give rise to reflexes. In the light of these considerations, the next section deals with goniometry.

4.4. Goniometry and Electromyography

One of the most important examinations of equilibrium function in otorhinolaryngology is goniometry (Fig. 4.6). The examinee is placed on a platform, which is gradually tilted, and tries to maintain the standing position in spite of the inclination. A normal healthy individual can maintain the standing position even when the platform is tilted about 25 to 30° in any direction, regardless of whether his eyes are open or closed. However, impairment of labyrinthine function results in marked disturbance of equilibrium in standing. This is particularly marked when the eyes are covered. Those whose labyrinthine function is entirely lost cannot stand on an inclination of even 2 to 5° and have to get off the platform. The standing position is probably maintained in spite of the goniometric incline by the coordinated actions of the muscles of the lower limbs. It has long been known that subjects with impaired labyrinthine function are incapable of maintaining the standing position. However, details of the mechanism of this impairment are not known and must be clarified by electromyography. I have not found any reports from abroad on this subject, but the following is a summary of some Japanese studies.

FIG 4.6. Goniometry.

Honjo and Furukawa examined the soleus muscle of normal and deaf subjects who were placed on an incline. They paid attention to the angular velocity of the inclination and varied it from 1–10° per second. At low angular velocities, there was no difference between normal and deaf subjects when their eyes were open, but there was a marked difference when their eyes were closed. At high velocities, there were differences between normal and deaf subjects even when their eyes were open.

My co-workers and I studied this problem for a year. We found that we could detect delicate movements in the lower limbs, particularly the toes and legs, when the subject was on the slope. The most marked changes were seen in the toes, which showed plantar flexion when the slope became steeper, behaving as if they were grasping the incline, and then abruptly becoming fully dorsiflexed to tolerate the slope. The standing position was maintained in spite of considerable inclination by alternate movements of agonists and antagonists. These movements were seen in healthy normal subjects.

As mentioned above, maintenance of the standing position was considerably disturbed in subjects whose labyrinthine function was impaired. In these subjects, we observed that antagonist muscles did not function or functioned in clumsy coordination. Therefore, we recorded electromyograms simultaneously from the gastrocnemius and the anterior tibial

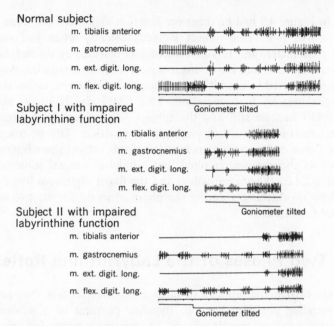

FIG. 4.7.   Electromyograms recorded simultaneously from antagonist muscles of leg and toes.

muscles, the antagonist muscles of the leg, and also from the extensor digitorum longus and the flexor digitorum longus muscles, the antagonist muscles of the toes. The results are shown in Figure 4.7. In the normal subject, as the goniometer incline is tilted backwards (making the subject lean backwards), the anterior muscles, i.e., the anterior tibial and extensor digitorum longus muscles, are activated to make the subject tolerate the inclination, whereas in the next movement their antagonists, the gastrocnemius and flexor digitorum longus muscles, are activated for control. In this way, the antagonist muscles work alternately and make standing possible in spite of considerable inclination, until the physical limit is reached, at which all the antagonist muscles are fully activated and stiffened and the subject falls off the slope. Results were different in subjects whose labyrinthine function was impaired. There are various kinds of labyrinthine impairment: some patients have completely lost their inner ear function and no nystagmus can be invoked; in others the functions of the two ears differ with apparent spontaneous nystagmus and Romberg's phenomenon, i.e., a Ménière's condition. Therefore, no common pattern could be distinguished, but we obtained the following results on individual subjects. Subject I with labyrinthine impairment shown in Figure 4.7 had no inner ear function and could not stand on an inclination of 2–3°. This subject swayed violently when just standing on the goniometer, and this movement was detected as discharges from the anterior tibialis and the extensor digitorum longus muscles. When the goniometer started to tilt, all the muscles gave discharges, rather than the antagonist muscles in cooperation. This EMG clearly shows how the lower limbs became stiff and the subject fell. Subject II in Figure 4.7 had unilateral impairment of labyrinthine function. The gastrocnemius and the flexor digitorum longus muscles of this subject show intermittent activities as the goniometer tilted, but, unlike in normal subjects, their antagonists, i.e., the anterior tibial and the flexor digitorum longus muscles, show no discharges during the phase when the former two muscles were not functioning.

4.5. Two Phases of the Labyrinthine Reflex

As stated earlier, the problem of posture involves reflexes. Not only the static standing position, but also dynamic postures of movements are the results of various muscle reflexes; i.e., posture is a manifest expression of the integration of various reflexes. There are various important reports on the changes in postures caused by muscle reflexes mediated by different receptors. For example, Takagi reported studies on the pressure on the skin; Tokizane, Morimoto, Honjo, Yoneda, Ino and we reported studies on the labyrinth, using a swing, rotating chair, or Grahe's goniometer; Morimoto and we reported on vision.

I have pointed out a strange aspect of the labyrinthine reflex. The labyrinth is responsible for very important equilibrium reflexes as shown by goniometric examination. Nystagmus is often examined in studies on labyrinthine function in which strong superphysiological stimuli are applied, such as ten rotations in 20 seconds or infusion of cold water into the ear. One aspect of labyrinthine function can be determined objectively by evaluating the resulting nystagmus. However, although such stimulation results in nystamus, the nystagmus occurs at the time of disequilibrium and motor ataxia when equilibrium is not maintained; that is, at the time of complete disruption of equilibrium when the subject is unable to remain standing and falls down, showing a severe kind of Romberg's phenomenon. Although the labyrinth controls equilibrium, it has the strange action of disrupting the equilibrium of muscles throughout the body and causing nystagmus when strong stimuli are given. Consequently, I proposed that the labyrinthine reflex should be classified into two phases: equilibration and its disruption caused by dif-

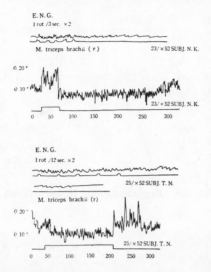

FIG. 4.8.   Two examples of phases in labyrinthine reflex caused by
rotations to the right at different speeds.

ferent degrees of stimulation. These two phases were demonstrated in
the stepping test and in the vertical writing test with eyes covered. The
tonus of the muscles in the phase of equilibration was opposite to that
in the phase of disrupted equilibrium. Morimoto and Ogino showed this
by electromyography as seen in Figure 4.8.

4.6. Dynamic Posture

Traditional studies of posture were directed to the static posture, particularly the standing position. Obviously standing is the fundamental
posture of man, but various complex movements are conducted on the
basis of this fundamental posture. The only studies made so far on dynamic postures have been those on changes in the location of the center
of gravity in various sports including running and analyses of motion
pictures of various athletic movements by Howorth (1946). Howorth supported the concept of dynamic posture, as shown in Figure 4.9, but he
only described the dynamic posture shown in Figure 1.38, which he
concluded was basic to all movements. Studies on this dynamic posture
are very important not only for sports, but also for efficient performance
of various jobs. However, the correct approach is necessary for inves-

FIG. 4.9.   Reproduced from Howorth (1946).

FIG. 4.10

tigating posture: simple moving pictures of changing postures do not lead to the heart of the problem.

I approached this problem from the standpoint of reflexes (see Chapters 1 and 2). First, by analyzing various forms of athletic movements, I showed that the neck reflex and the labyrinthine reflex are fundamental to human postures of movement. I also showed that Babinski's reflex and the supporting reflex (*Stützreaktion*) are fundamental to human dynamic postures. Basic to this idea that posture is supported by reflexes is the objective proof that these reflexes are latently present in the muscles of healthy normal subjects, as shown by electromyography. Tokizane,

FIG. 4.11. Reproduced from Tokizane *et al.* (1951).

Shimamoto *et al.*, and I showed by electromyography that the neck reflex, Babinski's reflex, and the supporting reflex (*Stützreaktion*), which had been considered to be pathologic reflexes appearing in subjects with neurological impairment, are present latently in normal subjects.

Figure 4.10 shows the form of a skier, which I will analyze from the standpoint of reflexes, including the lumbar reflex. The skier's form is splendid: the trunk is bent strongly to the left at the waist. Tokizane *et al.* demonstrated the lumbar reflex in healthy subjects by electromyography (Fig. 4.11). When the trunk is tilted relative to the lower limbs, the upper limb on the tilted side is flexed, whereas the other limb is extended. The lumbar reflex in a normal subject can also be seen in the skier's form. The maximum muscle force can be attained when turning at the maximum speed by assuming a form consistent with the lumbar reflex. In other words, manifestation of the lumbar reflex in the muscles, in addition to being beautiful, gives the maximum turning speed. In this way, dynamic posture should be interpreted in terms of reflexes. The lumber reflex is not the only reflex involved in this form. With the trunk tilted to the left, the head is moved to the right and thereby aligned with the line of the center of gravity. This is a labyrinthine and optic righting reflex, which makes the head turn to the right relative to the trunk. This evokes the neck reflex in the upper limbs, which strengthens the extension of the right upper arm and flexion of the left upper arm. The right lower limb is flexed more, which is consistent with the lumbar reflex which extends the lower limb on the tilted side and flexes the other limb. The strong flexion of the lower limb on the right side, i.e., the tilted side in skiing, is a result of the lumbar reflex. Therefore, at this instant, the lumbar reflex, the labyrinthine and optic righting reflexes, and the neck reflex are all functioning. In this way, dynamic posture can be clearly understood in terms of reflexes. Electromyography has contributed a great deal to this concept by showing that these reflexes, which had been considered to be pathological, actually exist latently in healthy normal subjects.

The subject of posture will be elaborated further by evaluating reflexes using electromyographic studies.

Part 2
Vestibular Function Evaluation

Chapter 5

Vertical Writing with Eyes Covered

5.1. A New Test of Vestibulospinal Reaction

It was in 1943 that the author reported in Japan a new test of vestibulo-spinal or deiterospinal reaction named *shagan-shoji-hō*. I devised this test when I studied the vestibular deviation phenomenon using the pointing test (Bárány, 1907a), *Abweichreaktion* (Güttich, 1940), *Armtonusreaktion* (Wodak and Fischer, 1923), etc.

This test has two advantages. First, it gives purely objective results on deviation or drifts, since the deviation is recorded on paper by the subject tested in the form of a line of letters. Second, it is very sensitive. These characteristics have enabled us to use it to find various new facts in the fields of labyrinthine physiology, diagnosis and treatment. To date, interesting results have been obtained by comparing findings before and after various newly developed operations on the ear; e.g., fenestra-tion, mobilization of the stapes, Portman's operation, and stellate block for Ménière's diseases.

However, the test, as well as these results, are scarcely known by in-vestigators in the rest of the world. In 1957, Wodak wrote that "in every vestibular examination, attention should be drawn not only to nystagmus but also to the deiterospinal reflex." Coincidence of opinion with him has stimulated me to publish this description of the test and a summary of the results obtained in the fields of labyrinthine physiology, diagnosis and treatment.

Method: The subject is seated on a chair facing a desk with his body upright and his eyes open. He is then made to write a vertical list of letters (which may be Roman letters, Chinese characters, Japanese *kana* letters, or marks) on a paper placed on the desk. Each of these letters is made to cover a space of 1×1 to 5×5 square cm, and a list of them is written vertically (from top to bottom on the paper), extending a total length of 5 to 20 cm. A pencil with a soft lead, or a felt-tipped pen, is

79

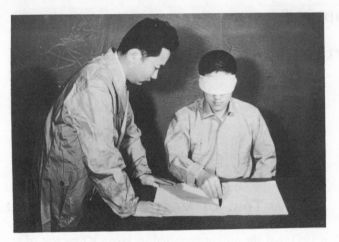

FIG 5.1.   Vertical writing test with eyes covered.

usually suitable for the purpose. The record thus obtained is entitled "letters with eyes open." After completing this procedure, the subject is blindfolded with a band or is asked to close his eyes, and writes the same list of letters vertically in the same scale, trying to keep them in a straight line. In this procedure, the examiner must give him a starting point by bringing the pencil in his dominant hand exactly in front of him and putting its tip on the paper, which must be fixed on the desk while the list of letters is being written (Fig. 5.1). During writing, the face of the subject must be directed exactly forward, and no part of his body should touch the desk; i.e., his chest, abdomen and upper extremities are all separated from the desk; the unused hand is placed on the knee, and the arm holding the pencil is maintained without any contact with the desk, so that the tip of the pencil is the only contact with the paper. The record thus obtained is entitled "letters with eyes covered."

The letters with eyes covered should be compared in their disposition and shape with those with eyes open. The disposition of the latter is straight and vertical with little, if any, deviation to either side in most persons except those with severe ataxia. When the letters with eyes covered show deviation in their disposition from one letter to another in one direction, right or left, the phenomenon is called "deviation to the right or to the left in vertical writing with eyes covered" or, briefly, "right or left deviation in writing." The angle of deviation is measured, drawing a line from the center of the first letter to that of the last letter or to that of the letter showing the greatest deviation from the first letter. A quantitative expression is then made, for example, as "left deviation in writing of 30 degrees." This test is indicated not only in cases of labyrinthine disturbances, but also in cases of cerebellar and other intracranial diseases. In these cases adiadochokinesis is expressed as derangement in a list of letters.

Writing with eyes covered should be repeated to obtain three to five serial records, while one record is usually satisfactory in writing with eyes open, which serves as a control. In this test a sectioned paper is a convenience in measuring the angle of deviation in writing. In our laboratory a sheet of paper 55×40 cm with 1.5-cm square sections is routinely used. The letters in the figures in this chapter are reduced in size from the original ones and the section lines are erased for the sake of clearness.

5.2. Letters with Eyes Covered in Normal Subjects

Sometimes a normal person with eyes open may show deviation in writing when no special attention is paid to avoiding deviation or when the posture is incorrect. It seems, therefore, necessary to examine practically whether or not normal persons can write precisely in a vertical direction without visual sense and in the posture described above, though letters with eyes open usually show no deviation in normal persons. Our examination of 200 adults with apparently normal labyrinthine function revealed the following facts:

1) Most of the persons tested could write "letters with eyes covered" in approximately the vertical direction. Deviations in disposition of letters were within the limit of 5 degrees on each side. Therefore, spontaneous deviation in writing, i.e., deviation observed without any stimulating procedure, within this limit should be considered normal.

2) When the deviation in writing is within 6 to 9 degrees on either side, imbalance of the vestibular function may be assumed. However, since some persons having no distinct signs of labyrinthine impairment showed deviation to this degree, that assumption may not always be correct.

3) Imbalance of vestibular function can be diagnosed with certainty when the spontaneous deviation in writing exceeds 10 degrees on either side. This degree of deviation could not be found in repeated writing tests on normal persons having no ear diseases.

5.3. Effects of Vestibular Stimulation on Letters with Eyes Covered

In testing vestibular function, special emphasis has hitherto been placed on the observation of nystagmus, which is routinely produced by rotatory or caloric stimulation. Tests for the vestibulospinal reflex or the deiterospinal reflex have been less important in labyrinthine examination. It is well known that lateralization in past-pointing, *Abweichreaktion*, etc. occurs on the side of the slow phase of nystagmus. This has also been proved in blindfold vertical writing. When nystagmus was produced by rotation or stimulation of the ear with cold water, the letters with eyes covered showed marked deviation in their disposition in the direction of the slow phase of nystagmus. To be emphasized in this connection is the fact that the writing showed marked deviation, and thus demonstrated functioning of the labyrinthine reflex, when rotatory or caloric stimulation were too weak to induce nystagmus. This sensitivity, a characteristic of the test, has already been referred to. The deviation in writing induced by such mild labyrinthine stimulation may be considered to be based on a quite different mechanism of physiological importance from that underlying production of nystagmus and deviation in writing during manifest nystagmus.

Results of testing in 100 normal subjects the effects of vestibular stimulation on letters with eyes covered are summarized in the following.

5.3.1. Rotatory Stimulation

A board is attached to a rotating chair so as to be located horizontally in front of the rotating subject who is made to write with his eyes covered successive vertical lists of letters on a paper placed on the board during and after rotation. The records thus obtained afford qualitative and quantitative evidence of functioning of the vestibulospinal reflex during and after rotation. The record during rotation is influenced by the centrifugal force produced by chair rotation. Since a considerably strong centrifugal force may be added to the arm holding the pencil in the case of intense chair rotation, deviation that may be observed in the record cannot always be regarded as of pure vestibular origin. How to interpret and analyze the data obtained during rotation is now under consideration. Therefore, only deviation in writing after rotation is recorded here. For the writing test after rotation alone, a board attached to the rotating chair is not always necessary; a desk may be sufficient if the rotating chair is stopped in such a position that the subject is precisely in front of the desk so that he can write immediately.

Postrotatory Deviation in Writing

a) *After 2 turns in 10 seconds*——A mild degree of rotation almost always induced postrotatory deviation in writing. The angle of deviation was between 5 and 30 degrees in most cases, but deviation up to 40 degrees was seen in some cases. The angle decreased gradually with lapse of time following completion of rotation, but definite deviation could still be observed in most cases 20 to 30 seconds after cessation of rotation. The deviation usually disappeared in about 60 seconds. The direction of postrotatory deviation was always left in the case of right rotation, and right in left rotation (Fig. 5.2). There were many cases in which much milder rotation, such as 1 turn in 5 seconds or 1 turn in 10 seconds, in duced distinct postrotatory deviation in writing.

b) *After 10 turns in 20 seconds*——This amount of rotation is commonly used in testing for postrotatory nystagmus. Following its completion, there was pronounced deviation in writing with eyes covered (Fig. 5.3). The angle of deviation ranged from 10 to 40 degrees in most cases, but sometimes it reached up to 60 degrees. The period during which deviation in writing was observed far outlasted that of postrotatory nystagmus, and many cases showed marked deviation in writing for as long as 60 seconds after cessation of rotation.

It is of special interest with this degree of rotation that the direction

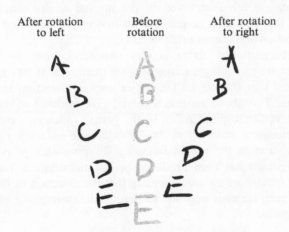

FIG. 5.2. Writing of normal subject before and after rotation (2 turns in 10 seconds) to the left and right.

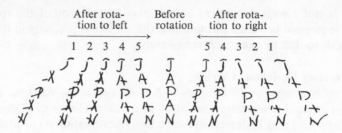

FIG. 5.3. Writing of normal subject before and after rotation (10 turns in 20 seconds) to the left and right. The columns were written serially (1–5) after cessation of rotation.

of rotation in writing is often reversed with the lapse of time following completion of rotation. Immediately following completion, when pronounced postrotatory nystagmus can be observed, the direction of deviation in writing usually occurs with the slow phase of the nystagmus. This fact can be explained by the results obtained by various methods, such as past-pointing and *Abweichreaktion*. The angle of deviation in writing usually decreases with the lapse of time until little deviation can be found. At about this time, postrotatory nystagmus usually stops. After some moments, deviation in writing, which has disappeared, generally becomes manifest again, but in the opposite direction this time. Then this deviation also becomes less marked until at last no deviation can be found. This phenomenon, deviation in writing in the direction opposite to the slow phase of postrotatory nystagmus following its disappearance, was often observed in the normal adults tested. The direction of this deviation coincides with that of the deviation in writing observed following 2 turns in 10 seconds.

The fact mentioned above is very important from the standpoint of labyrinthine physiology, diagnosis and treatment. It has been generally accepted that the labyrinthine reflex does not function until nystagmus occurs. The above results, however, give evidence of the functioning of the vestibulospinal reflex in the period following completion of mild rotation—so mild as to be incapable of inducing postrotatory nystagmus—and in the period following disappearance of postrotatory nystagmus which has been produced by optimal rotation. Further, it is a new and heretofore unknown finding that the direction of deviation in writing in such states is opposite to that of the slow phase of postrotatory nystagmus.

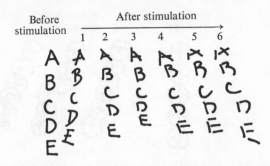

FIG. 5.4.   Writing by normal subject before and after infusion of
10 cc of water at 20°C into the right ear.

5.3.2. Caloric Stimulation

When vertical writing with eyes covered was examined following infusion of water into one ear, deviation occurred in the direction of the slow phase of nystagmus concurrent with manifestation of nystagmus, as seen in other tests on the vestibulospinal reflex. Though not constantly, a finding similar to that seen following rotation was sometimes observed in cases of caloric stimulation; i.e., deviation in writing in the direction opposite to the slow phase of nystagmus was sometimes observed in the period preceding appearance of manifest nystagmus (the latent period of nystagmus) and in the period following cessation of nystagmus (Fig. 5.4).

Deviation in writing which is induced by rotatory as well as caloric stimulation has been described so far—kinds of stimulation commonly used in examination of nystagmus. Worthy of note are rare cases of healthy ears in which changes in pressure in one external auditory meatus induced deviation in writing with eyes covered. The deviation was found to be reversed in direction according to increase or decrease in the pressure. In these cases, the so-called fistula symptom could not be observed.

5.4. Letters with Eyes Covered in Cases of Various Ear Diseases

Examination of a number of patients suffering from various ear diseases revealed that many patients with ear diseases showed marked spontaneous deviation in writing with eyes covered without any stimulation. It was also found in examining groups of healthy middle-school students

Written with   Written with
eyes open   eyes covered

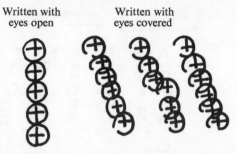

FIG. 5.5.   A case of supine positional nystagmus (30-year-old male).
The column of marks written with eyes open and with
head in the normal position (far left) is precisely vertical.
But columns of the same marks written in the supine
position with eyes covered show marked deviation owing
to positional eye nystagmus to the left.

Written with   Written with
eyes open   eyes covered

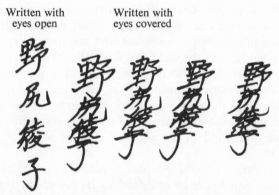

FIG. 5.6.   A case of the interval period of Ménière's disease (28-year-
old female). The column of Chinese characters (the patient's name) written with eyes open (far left) shows no deviation, whereas the other four columns written with eyes
covered show left deviation.

who could be regarded as having normal labyrinthine function that some
individuals showed marked spontaneous deviation in writing, and most of
them were found to have had ear diseases past or present.

Results of examination of patients with various ear diseases are summarized as follows:

1) In cases of vertigo of presumable vestibular origin, the letters with
eyes covered usually showed marked spontaneous deviation in one constant direction with a roughly constant angle in serial records (Fig. 5.5).

During
sickness

After
cure

FIG. 5.7.   A case of left acute suppurative otitis media (21-year-old male). The column of Chinese characters (the date) written with eyes covered during sickness shows left deviation of about 12 degrees; deviation is absent in that written two months later, after cure.

Before
treatment

After removal
of exudate
by puncture

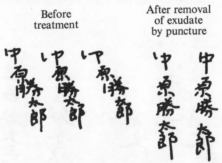

FIG. 5.8.   A case of right exudative otitis media (45-year-old male). Three versions of the patient's name written with eyes covered before treatment (left) show right deviation of 20–25 degrees, but deviation disappears after removal of exudate by puncture (right).

The deviation could be found not only in the period during which spontaneous nystagmus could be observed, but also in the period when the equilibrating function was considered normal, as in the interval period of Ménière's disease (Fig. 5.6). When deviation in writing was observed during manifestation of nystagmus, the direction of the former coincided with that of the slow phase of the latter, or in some cases with that of the rapid phase.

2) Patients suffering from various ear diseases that caused neither vertigo subjectively nor distinct disturbance of equilibrium objectively, e.g. patients suffering from acute or chronic otitis media or having undergone various ear operations, often showed marked spontaneous deviation

Before treatment   After catheterization

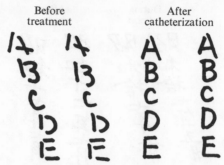

FIG. 5.9.   A case of obstruction of the right auditory tube (32-year-old male). Two columns of letters written with eyes covered before treatment (left) show right deviation, but deviation disappears after catheterization of the right auditory tube (right). (As can be seen in these figures, the inclination of deviation in the writing test is not always in a precisely straight line but sometimes curves.)

Before treatment   After catheterization

FIG. 5.10.   A case of obstruction of the left auditory tube (50-year-old male). Two versions of the patient's name in Chinese characters written with eyes covered before treatment (left) show left deviation of about 20 degrees, whereas deviation disappears after catheterization of the left auditory tube (right).

in writing in one constant direction with a roughly constant angle in serial records. This fact indicates the presence of imbalance in the vestibular function in such cases. It is noteworthy that definite spontaneous deviation of letters in one constant direction throughout serial records was often observed in cases of relatively mild noncomplicated middle ear diseases, e.g., otitis media purulenta acuta, otitis media exudativa, and obstruction of the auditory tube. Such deviation in writing was usually replaced by precise vertical writing or even new deviation to the opposite side when the otitis media had been cured, the exudate had been removed by puncture, or the auditory tube had been catheterized (Figs. 5.7–5.10).

5.5. Discussion

The fact that the letters with eyes covered show deviation when rotatory or caloric stimulation is applied to the labyrinth can be explained as a result of changes in tonus of the upper extremity musculature, which are caused by impulses originating from the labyrinth and passing through the vestibulo-spinal tract. The test of vertical writing with eyes covered is of much value in that it can offer definite objective evidence of such changes in muscle tonus even when these are slight and obscure subjectively as well as objectively. In comparing nystagmus and deviation in writing, the latter can react to far feebler stimuli applied to the labyrinth than the former. Examples have shown that deviation in writing can be observed following 2 turns in 10 seconds, when nystagmus is almost absent, as well as in the period following disappearance of postrotatory nystagmus and vertigo induced by 10 turns in 20 seconds. Further, the deviation in writing in the latter period is essentially different from that observed in the period of manifest nystagmus in that the direction of deviation is opposite to the slow phase of nystagmus, contrary to the direction of deviation during manifestation of nystagmus. The direction of deviation in writing following 2 turns in 10 seconds is also opposite to the slow phase of nystagmus which might be observed should the rotation be made optimal in the same direction. These results have revealed that there are two phases—characterized by contrary directions of deviation in writing—in the labyrinthine reflex, one seen during manifestation of nystagmus and the other seen in cases of labyrinthine stimulation so weak as to be incapable of inducing nystagmus. These two phases can often—but not always—be differentiated also in experimentation with cold water. These facts indicate the necessity of reconsidering the theory that the labyrinth is stimulated and then the labyrinthine reflex is elicited at the moment when nystagmus appears. They show that the vestibulo-spinal reflex can be evoked in man by stimuli to the labyrinth so weak that they cannot induce nystagmus, but which yet produce changes in tonus of the entire body musculature.

Patients suffering from various ear diseases, especially noncomplicated middle ear diseases, showed marked deviation in writing which could not be observed in persons with healthy ears. This fact may be explained as follows. Inflammations or pressure changes in one middle ear may duly influence the labyrinthine function of that side through the round window and the oval window, causing imbalance of the function between the two sides. This may result further in imbalance of the tonus of the entire body musculature between the two sides, which is under the control of vestibular impulses. When the imbalance is marked, objective symptoms, such as spontaneous nystagmus, staggering gait, or falling while standing (Romberg's phenomenon), as well as subjective sensation of vertigo, may appear. However, when the imbalance is slight, changes in muscle tonus may remain latent, being adjusted and compensated unconsciously by visual sense, and no distinct symptom may appear. This slight imbalance in labyrinthine function can be clearly demonstrated by spontaneous deviation in writing. In strong support of these considerations are the above-described cases of unilateral otitis media exudativa, in which deviation in writing disappeared after removal of exudate by puncture, as well as cases of unilateral obstruction of an auditory tube, in which catheterization caused disappearance of deviation in writing.

In conclusion, the vestibular labyrinth has been found to react to very slight physiological stimuli or pathological conditions which had hitherto been thought insignificant. To date the state of slight labyrinthine imbalance referred to here has been vaguely described under the name of latent nystagmus, *Nystagmusbereitshaft* or *Nystagmusneigung*. The blindfold vertical writing test reported here has offered clear and purely objective evidence of the presence of that state.

Chapter 6

Diagnostic Significance of Letters Written with Eyes Covered, with Reference to Writing in Cerebellar Ataxia

6.1. Writing with Eyes Covered

In the blindfold writing test proposed by Hoshino and Fukuda, the subject is asked to write vertically several signs such as circles, triangles, crosses, Japanese letters, or Chinese characters. The qualitative and quantitative changes that appear in the course of testing in the arrangement among the components of each symbol are evaluated. In vestibular labyrinthine disturbance or when the labyrinth is experimentally stimulated, the components of letters written by such a subject will be shifted with regularity in a certain direction in the course of writing in such a way that the axial line drawn by connecting the centers of individual letters will deviate in a certain direction. This is termed labyrinthine deviation of writing (Fukuda, 1959a). With this method, not only the presence or absence of labyrinthine dysfunction but also its intensity and nature can be objectively evaluated. This new test is also significant and valuable in that extremely mild labyrinthine dysfunction, which has hitherto often been overlooked, can be detected.

We administered this writing test with eyes covered to patients with orthostatic ataxia with acoustic tumor, cerebellopontine angle tumor, and cerebellar tumor, and compared the letters written by these patients. Characteristic findings were observed in each of these diseases, and this test was revealed to be useful for differential diagnosis of these ataxias.

The subjects of this investigation were two patients with acoustic tumors, three with cerebellopontine angle tumors, and three with cerebellar tumors who were admitted to the Departments of Otorhinolaryngology and Surgery, the Kyoto Imperial University Hospital. Their clinical findings and operative procedures (summaries of the records from the Surgery Department) are outlined in Table 2.1.

All eight patients performed the writing test with eyes covered and

TABLE 2.1. Clinical Findings and Operative Procedures for Subjects Tested

Case No.	Age and sex	Diagnosis	Operation	Neurological symptoms	Otoneurological symptoms	Illustrated writing
1.	24-year-old male	Acoustic tumor on the left side	A yellow-red, walnut-sized tumor originating from the left acoustic nerve was removed.	Bilateral papilledema	Subjective: vertigo, left-sided hearing impairment and gait disturbance. Objective: marked hearing impairment in the left ear. Normal hearing in the right ear. Hesitant in the stepping test. Nystagmus on bilateral gazing. Romberg's sign. Bárány's rotation produced nystagmus of 5″ with rightward rotation and 15″ with leftward rotation.	Fig. 6.1
2.	28-year-old female	Acoustic tumor on the left side	A smooth, almond-sized tumor (neurinoma) originating from the left acoustic nerve was removed.	No symptoms other than nerve disturbance.	Subjective: vertigo, hearing impairment and tinnitus in the left ear. Objective: marked hearing impairment of low frequency range in the left ear. Normal hearing in the right ear. Horizontal nystagmus with rightward gazing. 50 steps produced 20-degree leftward deviation. Romberg's sign. Falls leftward. Bárány's rotation produced nystagmus of 7″ with rightward rotation and 8″ with leftward rotation.	Fig. 6.2
3.	17-year-old female	Cerebello-pontine angle tumor on the left side	An almond-sized tumor (neurinoma) in the left cerebello-pontine angle was removed.	Bilateral visual impairment. Corneal reflex lost on the left side. Paresis of the left abducens nerve. Adiadochokinesis on the left side.	Subjective: hearing impairment and tinnitus in the left ear. Objective: marked hearing impairment in the left ear in high and low frequency ranges. Normal hearing in the right ear. Spontaneous horizontal nystagmus on the left side. Romberg's sign. Deviated 50 degrees to the left in the stepping test. Bárány's rotation produced nystagmus of 12″ with rightward rotation and 27″ with leftward rotation.	Fig. 6.3

4.	34-year-old female	Cerebello-pontine tumor on the left side	A pigeon-egg-sized tumor in the left cerebellopontine angle was removed.	Papilledema. Paralysis of all the branches of the left trigeminal nerve.	Subjective: hearing impairment on the left side. Objective: marked hearing impairment in the left ear. Normal hearing on the right side. Nystagmus on leftward gaze. No Romberg's sign. Bárány's rotation produced nystagmus of 10″ with rightward rotation and 20″ with leftward rotation.	Fig. 6.4
5.	51-year-old male	Cerebello-pontine angle tumor on the right side	Neurinoma originating from the right cerebellopontine angle was removed.	Hypesthesia in the first and second branches of the right trigeminal nerve. Right-facial paralysis.	Subjective: hearing impairment on the right side, tinnitus, gait disturbance. Objective: marked hearing impairment in the right ear. Normal hearing in the left ear. Nystagmus on leftward gaze. Deviated 40 degrees to the right in the stepping test. No Romberg's sign. Bárány's rotation produced nystagmus of 17″ with rightward rotation and 15″ with leftward rotation.	Fig. 6.5
6.	12-year-old female	Cerebellar tumor	An almond-sized tumor in the right cerebellar hemisphere was removed.	Visual disturbance in the right eye. Paresis of the right abducens nerve. Hyperesthesia of the second and third branches of the trigeminal nerve. Papilledema. Ataxia in the finger-to-nose test.	Subjective: tinnitus and hearing impairment in the right ear, difficulty in gait. Objective: marked hearing impairment in the right ear. No spontaneous nystagmus. Hesitant in the stepping test. Romberg's rotation produced nystagmus of 25″ with rightward rotation and 10″ with leftward rotation.	Fig. 6.6

(Continued on next page)

TABLE 2.1. (Continued)

Case No.	Age and sex	Diagnosis	Operation	Neurological symptoms	Otoneurological symptoms	Illustrated writing
7.	11-year-old male	Cerebellar tumor	Astrocytoma of a goose-egg size originating in the cerebellar vermis was removed.	Visual impairment, atrophy of optic disc. Reduced pharyngeal reflex. Leftward deviation of palatine velum. Disturbance in finger-to-nose and finger-to-finger tests.	Subjective: vertigo and difficulty in gait. Objective: bilaterally normal hearing. Nystagmus on bilateral gazes. Romberg's sign. Hesitant in the stepping test. Bárány's rotation produced nystagmus of 5″ with rightward rotation and no nystagmus with leftward rotation.	Fig. 6.7
8.	17-year-old male	Cerebellar tumor	Astrocytoma in the form of a cyst in the cerebellar vermis was removed.	Papilledema. Paresis of the right abducens nerve. Reflexes were exaggerated on the whole.	Subjective: tinnitus on the left side, gait disturbance. Objective: no hearing impairment on either side. Horizontal nystagmus with bilateral gazes. Romberg's sign. Bárány's rotation produced nystagmus of 20″ with rightward rotation and 18″ with leftward rotation.	Fig. 6.8

wrote their name or the date five times vertically. Comparative evaluation was made of the relationship of position and size among the written letters, angle of deviation of the axial line drawn by connecting the centers of individual letters, time lag of the components of each letter, and change in the nature and length of the components of each letter. The actual writings of the patients are illustrated (Figs. 6.1–6.8). Comments will be made on the characteristic findings in each of the diseases.

6.2. Acoustic Tumor

The patients wrote correctly when their eyes were open, whereas with eyes closed, the letters as a whole deviated in a certain direction either to the right or left. Each letter was written either similarly to the ones written at the time when the eyes were open (Fig. 6.1, Case 1), or deviated in a certain direction in the course of writing, while the relationship of position and length was retained between the left-side and right-side radicals or between the upper and lower radicals. In the latter situation, each letter was deformed in a certain direction, and the letters as a group deviated in the same direction (Fig. 6.2, Case 2). In the case

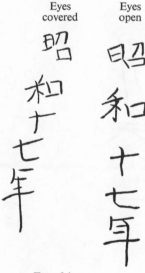

Eyes covered   Eyes open

FIG. 6.1

A subject (Case 1) with acoustic tumor wrote the date (昭和十七年) with eyes covered (left) and eyes open (right).

FIG. 6.2

A subject (Case 2) with acoustic tumor wrote her name (利斎ノブ) this way.

of acoustic tumor, the letters written with eyes covered were typically deviated as mentioned with reference to labyrinthine dysfuction.

6.3. Cerebellopontine Angle Tumor

The letters as a group deviate in a certain direction (Fig. 6.3, Case 3; Fig. 6.4, Case 4; and Fig. 6.5, Case 5). The angles of deviation and sizes of individual letters and spaces between succeeding letters are irregular. The letters either become progressively larger (Fig. 6.4) or smaller (Fig. 6.3). The space between the adjacent letters either becomes smaller, so that the next letter is written over the first letter (Fig. 6.4), or the space becomes wider (Fig. 6.3). Individual strokes of a letter deviate in a certain direction to different degrees in the course of writing, so that succeeding strokes are written over first strokes, or strokes become dispersed. In Figure 6.4, the third character is deformed in such a way that the right and left radicals are overlapping. Conversely, in Figure 6.3 the first and third letters are deformed in such a way that the right and left radicals are dispersed. The proportion of the right and left radicals are unharmonious in these patients.

In cerebellopontine angle tumors, the written letters as a group de-

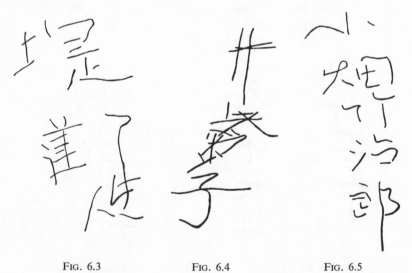

FIG. 6.3   FIG. 6.4   FIG. 6.5

Writing of subjects (Cases 3–5) with cerebellopontine angle tumors. The characters are the patients' names: Fig. 6.3, 堤道枝; Fig. 6.4, 井上鈴子; Fig. 6.5, 小畑竹治郎.

viate in a certain direction. In addition, the strokes of an individual letter deviate with irregular deviation angles and varying relationships of length and position.

6.4. Cerebellar Tumor

When the patients were asked to write a series of letters five times vertically, no deviation in a certain direction was observed. The letters deviated slightly to the right or left, presenting no uniformity. The angle of deviation is within the normal range of 10 degrees or less. The angle of deviation and size of individual letters and the spaces between them are irregular and disharmonious. In Figures 6.6 (Case 6) and 6.7 right (Case 7), the first and second letters are spaced too wide, whereas the second and third letters are overlapping. In Figure 6.8 (Case 8), the first letter is smaller than the second, whereas the second and third letters are of the same size. In Figure 6.7 left, these relationships are reversed. In each of the individual letters, their components are unprincipled and unbalanced in terms of length, arrangement, or position. In the first letter of Figure 6.6 and the third letter of Figure 6.7, the right and left

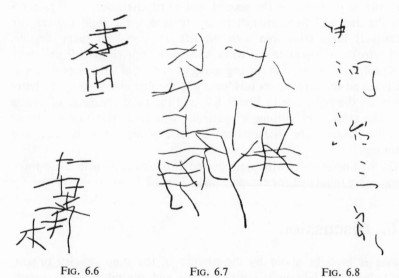

FIG. 6.6   FIG. 6.7   FIG. 6.8

Writing of subjects (Cases 6–8) with cerebellar tumors. All columns of characters except the right one in Fig. 6.7 are the patients' names: Fig. 6.6, 稲垣梢; Fig. 6.7 left, オチユー一郎; Fig. 6.7 right, 大日本; Fig. 6.8, 中河治三郎.

TABLE 2.2. Comparison of Writing with Eyes Covered by Patients
with Various Brain Tumors

		Acoustic tumor	Cerebello-pontine tumor	Cerebellar tumor
Deviation	qualitative*	regular	regular	irregular
	quantitative**	regular	irregular	irregular
Distance between and arrangement of individual letters and their strokes		harmonious***	inharmonious	inharmonious

* Deviation is qualitatively regular when it is in a certain direction, that
is, if the axial line or curve drawn by connecting the centers of characters
is of a certain slope or of a certain tangential angle.

** Deviation is quantitatively regular when each stroke or character deviates
with a more or less certain angle. When these angles are varied, the letters
deviate with quantitative irregularity. The group of letters with qualitative
and quantitative regularities is arranged so that their axial line is straight
and sloped.

*** Letters are arranged harmoniously when the size and mutual distance of each
character and the length and arrangement of each stroke are balanced, even
when they deviate.

radicals or upper and lower radicals are overlapping. In the first and
third letters of Figure 6.6, the right and left radicals are shifted from
the normal position. In the second and third characters of Figure 6.8
and the first and third characters of Figure 6.6, the right radicals are
excessively large compared with the left ones, making them illegible.
Each stroke is found to be wavy and unsmooth (Figs. 6.7 and 6.8).
Unsteadiness is marked at the beginning or ending of strokes as shown in
the first and fifth characters of Figure 6.8, the first character of the letter
group on the right side in Figure 6.7, and the third character of Figure
6.6. The strokes of writing by patients with cerebellar tumors whose
eyes are covered are characterized by lack of regularity, balance, and
harmony.

The writing samples with eyes covered of the patients with these three
diseases are tabulated for comparison in Table 2.2.

6.5. Discussion

Ataxia is brought about by disturbance in the deep sensory organs,
vestibulum and labyrinth, visual organs, and central nervous system,
which are in charge of the coordination of bodily movements. Depending
on the site of disturbance, ataxia is classified into peripheral, spinal
cord, cerebellar vestibular, brain-stem, and cerebral. Various methods

TABLE 2.3. Comparison of Labyrinthine and Cerebellar Ataxias

	Labyrinthine ataxia	Cerebellar ataxia
Clinical course	Mild. Alleviated in the course of time	Marked. Not alleviated in the course of time
Spontaneous nystagmus	Regular	Irregular
Hearing impairment	Present	Usually absent
Standing and gait	Falls or deviates in a certain direction	No certain directionality in falling or walking
Adiadochokinesis	Absent	Present
Fine complex movement	Undisturbed	Disturbed
Spontaneous deviation	Bilateral	Unilateral
Speech disturbance	Absent	Present
Vestibular labyrinthine response	Abnormal	Mostly unchanged

of differential diagnosis have so far been proposed, but they are complicated and inaccurate. Cerebellar and labyrinthine ataxias, which are differentiated in the field of otolaryngology, are the most difficult. These two ataxias are differentiated as in Table 2.3, according to the existing criterion.

The blindfold writing test will give the following results in labyrinthine ataxia. The size of each letter and their mutual distance are relatively accurate and balanced. The letters as a group deviate in a certain direction, and the axial line drawn by connecting the centers of individual letters is slanted in a certain direction. Left and right radicals or upper and lower radicals of a character are shifted in a certain direction, so that the right side of the character becomes either higher or lower than the left, or the upper part of one character is situated more leftward or rightward than normal. These are the characteristics of what is called labyrinthine deviation of writing. In cerebellar ataxia, the letters as a group stay undeviated or become deviated within the range of deviation angle for the normal subjects. The size of individual letters and their spacing are uneven and irregular. The length, direction, arrangement, and spacing of strokes of a letter are qualitatively and quantitatively nonuniform, disharmonious, and unsmooth. The right and left radicals or upper and lower radicals are out of their appropriate positions and form illegible characters. These letters are conspicuous and unique and are called the writing of cerebellar ataxia.

The mechanism causing the difference in the results of the blindfold writing test between labyrinthine and cerebellar ataxias is considered to be the following. Labyrinthine ataxia is due to disturbed balance of skeletal muscular tensions that are governed by the labyrinth. In this type of ataxia, movements do take place smoothly, but they do not proceed pertinently and deviate in a certain direction, for there is dysequilibrium of the tension in antagonist muscles concerned with the movements. In writing with eyes covered, by the same token, the group of letters as well as individual strokes are shifted in a certain direction, while retaining harmony.

In cerebellar ataxia, there is disturbance of skeletal muscles so that associated movements are disturbed and diadochokinesis develops in addition to ataxia. In this condition, although simple movements are possible, associated movements which are complicated and made up of contraction and relaxation of several muscles are not performed accurately or smoothly. In writing of symbols or characters, which is a complicated associated movement, continuation and conversion are uncoordinated so that each stroke of characters becomes qualitatively and quantitatively disfigured to produce the unique appearances described above.

In this way, labyrinthine and cerebellar ataxias can be differentiated by the results of the blindfold writing test, owing to their characteristic mechanisms of development.

Chapter 7

Labyrinthine Ataxia and Cerebellar Ataxia

7.1. Introduction

In textbooks, labyrinthine ataxia (motor ataxia) and cerebellar ataxia
are often very clearly distinguished as if they were of completely different
nature and as if the labyrinth and the cerebellum performed different
functions. In clinical practice, however, only a few cases can be as clearly
differentiated as is described in textbooks, and we are usually faced with
difficulties in judgment. For example, even when there are abnormal
findings in the cerebellum on X-ray and other examinations, typical
labyrinthine ataxia can be present, and this condition can even be cured
by an operation on the cerebellum in some cases. Every time one encoun-
ters such cases, one feels the necessity for having a precise understanding
of anatomy in making a diagnosis. The labyrinth and the cerebellum do
not perform different functions but are in close developmental relationship. The cerebellum initially originated as the center of the vestibular
nerve (funicular nerve) and can even be called the vestibular brain.

The cerebellum is made up of this center of the vestibular nerve, which
communicates with the sensory spinal nerves from the periphery, cerebral
cortex, pontine nucleus, olivary nucleus, and other nerve centers. For
this reason, in cerebellar disturbance, labyrinthine ataxia can take place
if the lesion is in the scala vestibuli, which is in particularly close relationship with the vestibular nerve. This is often encountered clinically. In
other words, labyrinthine ataxia can develop as a form of cerebellar
ataxia. In textbooks, adiadochokinesis, walking with a zigzag pattern,
and rebounding phenomenon are listed without any system as typical
symptoms of cerebellar ataxia. What one has to know here is that
the cerebellum is made up of the vermis, cerebellar hemispheres, scala
vestibuli, scala spinale or neocerebellum, and paleocerebellum. The
function of each of these cerebellar components is as complicated as
that of the cerebrum. Bárány described which sites on the cerebellar

hemisphere were concerned with the change in the tension of which skeletal muscles, but the essence of its functions has not clearly been confirmed. Whatever the situation, people have been trying to classify cerebellar and labyrinthine dysequilibria and ataxias in order to make distinctions among dysequilibria caused by disturbance in the movement and tension of skeletal muscles. In the following, I will try to give my own solution to this problem on the basis of my own experience.

7.2. Two Types of Labyrinthine Ataxia

Labyrinthine ataxia is usually treated as a single entity, and it is overlooked that there are two types of labyrinthine ataxia. The labyrinth is an organ of bodily equilibrium located in the inner ear. It is composed of the semicircular canals and vestibular organs. The labyrinth is stimulated by gravity as well as by linear or circular movement of the body. These stimuli are transmitted via the vestibular nerve to the vestibular center of the medulla oblongata, and are then communicated to the muscles of the body by way of, or with no relation to, the cerebellum. The stimulation affects the ocular muscles, all the skeletal muscles, and smooth muscles. The muscle reflex produced by labyrinthine stimulation is distinguished from the ocular reflex, skeletal muscle reflex, and autonomic nerve reflex because it is a pure reflex without ascending fibers from the medulla oblongata. Therefore, in examination of labyrinthine function, the change is evaluated in the movement of ocular and skeletal muscles which are connected with each other in the reflex arc. The labyrinthine function is compared to a balance in the sense that the activated labyrinths on both sides constantly pull muscles with equal strength. If the function on one side becomes stronger or weaker than that on the other, the muscles are pulled more toward one direction as in a balance. When this occurs to skeletal muscles, a person falls in one direction or shows a positive Romberg's sign in one direction. When he is made to walk with eyes closed, he walks sideways in one direction and shows nystagmus in the eyes. These clinical findings are often encountered in otitis interna or an acoustic tumor of one side. Even healthy subjects become dizzy and unable to stay standing when the labyrinth is strongly stimulated by 10 rotations in 20 seconds on a chair, being spun while standing, or having cold water injected into an ear. In these instances, they fall in one direction or their walking deviates in a certain direction. This is called "deviation" and appears not only in the lower limbs, but also in the muscles of the upper limbs and throughout the body, as will be mentioned later. This labyrinthine ataxia of deviation has been well described.

There is, however, another form of ataxia which should be emphasized here. This other form appears when the labyrinthine function is destroyed bilaterally. No deviation occurs in this case, but specific symptoms are demonstrated. In streptomycin intoxication, the vestibular nervous systems are affected bilaterally, and, in the end, both sides become afunctional. In this instance, no one-sided deviation takes place, or one sees no deviation phenomenon, which is usually understood to be the labyrinthine ataxia itself. It should be remembered that there are two types of labyrinthine ataxia, one being the form caused by the difference in the functions of the two sides (type 1) and the other being the bilateral loss of function (type 2).

7.3. Outline of Cerebellar Ataxia

The labyrinth is a peripheral organ, and its reflex can be examined by applying various stimuli. The cerebellum, however, is an intracranial center concerned with involuntary control of skeletal muscles and cannot as easily be stimulated. Cerebellar ataxia in humans is therefore studied clinically and pathologically rather than experimentally. In animal experimentation, decerebellation and electric or thermal stimulation of the cerebellar surface are performed and the results are compared with human cerebellar ataxia. Cerebellar disturbance in humans is therefore important and should be examined in detail and evaluated in relation to operative and autopsy findings. Adiadochokinesis, or inability to perform rapid alternating movements such as rapidly rotating the open hands with the arms outstretched, is described as a necessary sign of cerebellar ataxia in some books. In the disturbance of the cerebellar vermis, however, the patient has strong ataxia and yet is often capable of diadochokinesis. No particular sites of the cerebellum have been confirmed to be related to specific symptoms (or such an assumption itself may be ungrounded). At present, one knows generally that there are different motor ataxias for the vermis and the hemisphere. The best example of cerebellar ataxia can be demonstrated in the growth of babies. After they have learned to walk, adults teach them the game of patty-cake in which babies learn to put palms together and to open them or to move their hands to touch their mouth. These are considered to be a practice of diadochokinesis. In kindergartens, children learn to dance with their hands up and to rotate them to simulate falling leaves. These can be considered to be training of the cerebellum. In babies' growth, the cerebellar function of the lower limbs is first established followed by the development of delicate movements of the upper limbs and fingers. The

FIG. 7.1.   A 17-year-old male subject with cerebellar ganglia tumor
wrote this version of his name, 壺井初成, with eyes
covered.

lower limbs are mainly controlled by the vermis, which is developmen-
tally older, whereas the hemispheres are newer and concerned with delicate movements of hands and fingers. The development of motor function
from newborn, to infant, to child corresponds with the maturation of the
cerebellar function. Cerebellar disturbance can generally be considered
to be a process reversing this. In cerebellar ataxia, adiadochokinesis,
dysmetria, hypermetria, and asynergia appear, which are summarized
as follows:
1)   Staggering gait, tottering, and drunken gait.
2)   Inability to rapidly stop walking.
3)   Upper body left behind when walking.
4)   Upper limbs not moving in combination with walking.
5)   Inability to sit up from the supine position without using the
      arms.
6)   Inability to move the hand(s) properly to a desired object.
7)   Passing by a desired object.
The above conditions are easily observed in the movements of infants
and children. In cerebellar disturbance of adults, various ataxias, which
are normally seen in childhood, are often present. The best example is
seen in written letters. For example, when asked to write letters or various
symbols, these patients write them with irregular shapes and inconstant
sizes as if written by children in kindergarten or the lower grades of primary school. Characteristic examples are given in Figure 7.1. They show
asynergia on dysmetria (hypermetria and hypometria). In contrast to
these, the examples in Figure 7.2 show deviation in writing produced in

FIG. 7.2.   Writing and marks which show characteristic deviation
produced in labyrinthine ataxia.

labyrinthine ataxia. Although there is deviation, there is a certain regularity, unlike in the examples in Figure 7.1.

Most textbooks enumerate more than ten symptoms of cerebellar ataxia. These numerous symptoms can be summarized, and cerebellar ataxia itself can be defined, as "reduction or absence of the reflex of the antagonist muscle(s) in performing a certain movement (adiadochokinesis and dysmetria), or lack of coincidently occurring movements when some part of the body moves in a certain way (synergia)." In labyrinthine ataxia, the muscles of the body are deviated in a certain direction; i.e., labyrinthine ataxia is different from cerebellar ataxia. This indicates a method of differentiating the two ataxias. Another important point of differentiation is that ataxia is manifested bilaterally in unilateral labyrinthine dysfunction, whereas in cerebellar ataxia, it appears on the side of the cerebellar lesion in principle.

Although one can differentiate an ataxia resulting from a lesion in the cerebellar vermis from one in the cerebellar hemisphere, no other determination can clearly be made of the location of the cerebellar lesion based on the clinical manifestation of ataxia. The disturbance in the vermis primarily causes ataxia in the lower limbs; in the upper limbs, ataxia of delicate movements of the hands and fingers is observed rarely. For example, even when the patient has such a marked ataxia from a disturbance of the vermis that he is incapable of walking without supporting himself by holding onto the wall or table, frequently he is capable of smooth diadochokinesis of the upper limbs. The gait of the patient with cerebellar ataxia is said in textbooks to be as if done while drunken, and the direction in which the patient falls is said not to be definite. However, the gait of those whose vermis is impaired is often such that first the head and then the trunk abruptly fall posteriorly. In cerebellar hemispheric impairment, however, adiadochokinesis and ataxia in the finger-to-finger or finger-to-nose test are the usual initial findings.

The physiology of the finger-to-finger or finger-to-nose test is interesting if one thinks about it, although we do not usually pay much attention to it when administering these tests. For example, lying in bed in the dark at night, one sometimes feels itchy and unconsciously brings one's finger exactly to the itchy place. One wakes up with an unusual sensation of one's finger pressing on a flea on the body. Responding to the itchy sensation, the finger is delivered as a reflex to one point in a wide area of the body where the flea is. This is a surprisingly delicate and exact reflex movement compared with the finger-to-finger or finger-to-nose test. Bárány (1907a) states that in the cerebellum there is a center of spatial orientation. In cerebellar disturbance, patients produce poor and inadequate results in the finger-to-finger and figer-to-nose test; that is, spatial orientation is disturbed. The presence of the appropriate reflex in spatial orientation, as indicated by the fact that even a flea can be pin-pointed during sleep, shows that there is normal physiological cerebellar function and that the cerebellum constitutes a center of reflex.

7.4. Vision and the Two Motor Ataxias

Another important difference between labyrinthine and cerebellar ataxias is said to be as follows. Labyrinthine ataxia is compensated for by vision, whereas such a compensation is hardly present in cerebellar ataxia. This differentiation is important and significant in clinical examination. Labyrinthine ataxia is said to be gradually alleviated, whereas cerebellar ataxia is not easily alleviated. This difference can also be explained by the fact that the former ataxia is gradually compensated for by vision, whereas such a compensation is absent in the latter. When one labyrinth is surgically removed or when, in animal experimentation, one labyrinth is injured or the vestibular nerve is severed, marked ataxia develops, which is reduced in the course of days. This is explained by visual compensation. In streptomycin intoxication, labyrinthine function is bilaterally reduced or destroyed. This ataxia is visually compensated for gradually, though not completely. This is indicated by the fact that marked ataxia manifests itself immediately following the closing of the eyes. For instance, a patient with streptomycin intoxication can raise one leg and maintain this posture for 30 seconds with his eyes open, whereas he cannot stand on both legs for even one minute with eyes closed. When testing the same patient standing on a horizontal beam by changing the slope of the beam in the anterior, posterior, leftward, or rightward direction to see at what angle he falls, the patient with labyrinthine ataxia can withstand an inclination to the same degree that normal subjects can with eyes open.

TABLE 2.4. Comparison of Tilting Tests in Patients with Labyrinthine and Cerebellar Ataxias

	(A) Labyrinthine ataxia				(B) Cerebellar ataxia			
	Forward tilt	Backward tilt	Left tilt	Right tilt	Forward tilt	Backward tilt	Left tilt	Right tilt
Eyes open	38°	32°	34°	37°	3°	4°	2°	2°
Eyes closed	3°	2°	2°	3°	2°	4°	2°	2°

However, once the eyes are closed, he falls from the beam even at 2° (Table 2.4, A). This clearly shows how vision compensates for labyrinthine ataxia.

In contrast, a patient with cerebellar ataxia can walk more or less normally using a cane. He falls, however, when the floor is tilted as little as 2°. This holds true regardless of whether his eyes are open or closed (Table 2.4, B). In other words, no visual compensation takes place.

Both these patients have normal visual acuity. Both of them possess normal vision, and there is visual compensation for labyrinthine ataxia, whereas no such compensation takes place for cerebellar ataxia. In the first patient labyrinthine function is bilaterally lost, which corresponds to type 2 labyrinthine ataxia mentioned earlier. With the eyes closed, the ataxia of skeletal muscles in this patient resembles cerebellar ataxia. In cerebellar ataxia, nystagmus can be experimentally elicited within a normal range. In the case of the first patient, however, nystagmus cannot be elicited, or is of very short duration. In addition, there is gradual visual compensation. Labyrinthine ataxia can thus be easily differentiated.

However, it is wrong to think that cerebellar ataxia cannot visually be compensated for at all. The proof for this is as follows. The finger-to-finger test and other clinical tests, which we perform daily without thinking much about, are administered with subjects closing their eyes, and based on the results of these tests, cerebellar ataxia is diagnosed. Patients with cerebellar ataxia can touch the examiner's fingertip or their own nose with their finger if their eyes are open, unless their ataxia is very marked. This shows that cerebellar ataxia can be compensated for visually. If there is no visual compensation in cerebellar ataxia at all, administering the finger-to-finger or finger-to-nose test with the eyes closed is completely insignificant. The labyrinthine ataxia or dizziness produced in normal subjects by rotating them many times is reduced when the eyes are closed. Labyrinthine ataxia in this case is reduced when vision is obstructed.

In spite of the above, it is important to realize that labyrinthine ataxia is easily compensated for, but it is unlikely that cerebellar ataxia is. This statement is not a rule, but a general principle. The question of vision is of particular importance in the case of bilateral reduction or absence of labyrinthine function. This impairment was mentioned previously as type 2, and its main manifestation is lack of deviation resulting from the difference in the functions of the two sides. In this type of ataxia, there is no deviation of skeletal muscles, that is, no pulling in one direction. This strongly resembles cerebellar ataxia. Walking, for example, is not deviated in one direction, but unsteady, as if drunken. Such a condition is encountered clinically in streptomycin intoxication and some deafness resulting from inner ear infection. This labyrinthine ataxia is characterized by its later compensation by vision. The condition improves to such an extent that ataxia is not felt by patients if they have their eyes open. Once the eyes are closed, however, marked motor ataxia develops, which is the feature of this dysfunction. This is easily differentiated from cerebellar ataxia as no experimental nystagmus develops after rotation or cold water stimulation.

It is obvious that vision is important for bodily equilibrium and movement. However, no clear-cut study was performed on the problem of the control of skeletal muscles by vision, although there is a vague concept of visual compensation. I raised the issue of vision in relation to the two kinds of ataxias in order to draw attention to research on vision and its effect on bodily equilibrium. The mutual relation among vision, labyrinth, and cerebellum should be studied.

7.5. Nystagmus and the Two Motor Ataxias

The issue of nystagmus in relation to the two ataxias is quite complicated and much remains to be clarified. One cannot simplify the problem, but, plainly stated, the following facts hold. When labyrinthine ataxia is a result of a difference in bilateral functions, spontaneous nystagmus is often encountered. There is also rotatory vertigo at the same time. Even when nystagmus is not observable, patients are in the preparatory state of nystagmus, and the smallest stimulation produces nystagmus. In the case of type 2 labyrinthine ataxia, i.e., bilateral reduction or absence of labyrinthine function, nystagmus and vertigo are encountered if the function is bilaterally lost gradually, as in streptomycin intoxication. In this case, the patient may go through a nystagmus preparatory stage. However, if the labyrinthine function is completely lost, nystagmus is not produced by, for example, ten rotations in 20 seconds or cold water infusion. In this case, even when nystagmus is produced, it is negligible or in the state of anystra or hyponystra.

Some textbooks state that in cerebellar disturbance, spontaneous nystagmus always develops and is strong and extensive. It is said that vertical nystagmus is often seen and gradually increases. Some books even say that such a nystagmus is the first indication of cerebellar disturbance. In this regard, I hold a very reserved viewpoint. In the past years, I conducted otological examinations on many patients with cerebellar disturbance (mostly tumors, but some otic abscesses). I found that only a small number of patients showed such a nystagmus. In some typical cerebellar tumors, I observed positional nystagmus and vertigo, which had to be interpreted as of labyrinthine origin. These tumors were deeply situated inside the cerebellum and were removed surgically resulting in complete disappearance of nystagmus and vertigo. In this case, the lesion was located in the vestibulum of the cerebellum and produced nystagmus and vertigo similar to those of labyrinthine origin. In short, spontaneous nystagmus does not have to be present in cerebellar ataxia, and nystagmus and vertigo that are usually interpreted as of labyrinthine origin can be observed. As is written in some textbooks, there is also spontaneous nystagmus of variable frequency, amplitude, and direction. One will not make mistakes if one keeps all three of these situations in mind.

7.6. Conclusion

As described above, the two ataxias can generally be differentiated. However, when, for example, an acoustic tumor grows large and presses against the cerebellum, purely labyrinthine ataxia develops first, followed by cerebellar ataxia, presenting a mixed form of ataxia. Precise distinction of this type of ataxia will be made elsewhere. It is emphasized that in clinical situations mixed forms of ataxia are not infrequent.

Chapter 8
The Stepping Test
Two Phases of the Labyrinthine Reflex

8.1. Introduction

Previously, the author reported a new test of labyrinthine function entitled "*shagan-shoji-hō*," or vertical writing with eyes covered (Fukuda, 1957). This test is a method of obtaining objective evidence of imbalance of labyrinthine function which is expressed by functional imbalance of the musculature of the dominant arm as deviation in writing. The stepping test to be reported here also aims to observe deviation that is induced by imbalance of the labyrinthine function, but is expressed mainly in the lower extremities during stepping.

Differing from vertical writing with eyes covered, which is an original method of the author, the stepping test is similar to that reported by Unterberger (1938) under the name of *Tretversuch*, as well as by Hirsch (1940) under the name of the waltzing test. The reason why I report here such a time-honored method under a new name as the stepping test is as follows.

At the time when these reports were made, I was investigating the labyrinthine deviation phenomenon in the upper extremities by means of the pointing test (Bárány) and *Armtonusreaktion* (Wodak and Fischer) as well as the vertical writing test which I had newly devised. Thus their reports profoundly interested me in that the test was concerned with labyrinthine deviation expressed in the lower extremities. The test then came to be used routinely in our clinic in parallel with the ordinary gait test, and these tests were compared with each other from the clinical as well as physiological point of view. The result revealed that *Tretversuch* or the waltzing test was so good that it could replace the ordinary gate test. But since the original method had some deficiencies, I modified and revised the test and introduced it in Japan under the name of the stepping test (*ashibumi-kensa*) some twenty-five years ago (Fukuda, 1957). Since then it has come to be widely used in Japan for testing spontaneous

110

deviation, and descriptions of the stepping test and the vertical writing test can be found in every Japanese textbook of otorhinolaryngology.

In contrast, it seems that in most other countries experimental eye nystagmus continues to be the principal means of testing labyrinthine function and that the stepping test (or *Tretversuch*, or the waltzing test) is not in general use. To be noted in this connection is the fact that both Unterberger and Hirsch seem to have placed primary emphasis on the deviation of the body during stepping (as well as on experimental eye nystagmus) induced by caloric labyrinthine stimulation (which can also induce experimental eye nystagmus) in entitling the phenomenon "*Körperdrehung um Langsaches*" or "spinning or waltzing."

From my own experience it has become clear that quite significant results can be obtained when the test is used for observing deviation of the body that occurrs spontaneously or without applying any experimental labyrinthine stimulation. For instance, affection of the labyrinth is suggested in this test in cases of unilateral acute otitis media or obstruction of the auditory tube without any sign of inner ear complications. The test can also reveal, by the appearance of marked deviation, the fact that such a mild degree of rotation as to be incapable of inducing eye nystagmus or changes in pressure in the external auditory meatus can stimulate the labyrinth to react. Thus the test can elucidate, by testing the lower extremities, various new facts in the field of labyrinthine physiology, diagnosis and treatment, similar to those that have already been clarified by means of the vertical writing test for the upper extremity.

In the following sections I will describe successively the procedure of the test and new facts obtained by means of the test in the field of labyrinthine physiology; and finally emphasis will be placed on the importance of observing the spontaneous labyrinthine deviation in the examination of labyrinthine function.

Unterberger's and Hirsch's Methods: Though Unterberger's test and Hirsch's test have some discrepancies in details, they have the same principal features; therefore these tests will be described here as one and the same test, which is carried out as follows.

The subject is asked to stand upright with his eyes covered and with both arms stretched straight before him and to flex and raise first one knee and then the other repeatedly. A normal individual can step in this manner in approximately the same position without any spinning movement of the body around its vertical axis. When eye nystagmus is induced by instillation of cold water into one external auditory meatus, the subject gradually spins around his vertical axis in the direction of the slow phase of eye nystagmus, though he is not aware of his spinning and believes he is stepping in the same position and facing in the same direction. About a 90° turn is induced by instillation of 10 cc of water at a temperature of 27°C, while about a 180° or complete turn is induced by instillation of 5 cc of water at 15°C. When the test is performed in a patient suffering from pathologic irritation of the labyrinth and showing spontaneous nystagmus, his body may spin spontaneously around his vertical axis in the direction of the slow phase of nystagmus.

On applying this test practically to many normal subjects and patients suffering from various ear diseases, I found it necessary to prescribe the total number of steps. Also I found that, while the body turned around its vertical axis, it might move from the original position forwards, backwards, or laterally. To obtain objective data regarding the nature and degree of the displacement during stepping, new devices have been made as described in the following section.

8.2. The Stepping Test

8.2.1. Modified Fukuda's Method

Two concentric circles having radii of 0.5 m and 1.0 m (or 4 concentric circles having radii of 0.5 m, 1.0 m, 2.0 m, and 3.0 m) are drawn on the floor. The circles should be divided into sections by lines passing through the center at 30° or 15° angles. The subject is asked to stand upright in the center of the circles with his feet close together.

FIG. 8.1

Fɪɢ. 8.2

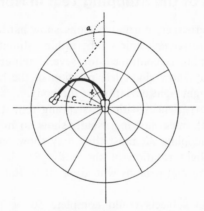

Fɪɢ. 8.3. Locus of stepping deviation (heavy black line). Roman
characters indicate (a) angle of rotation, (b) angle of
displacement, and (c) distance of displacement.

The subject is then blindfolded with a band and asked to stretch both
arms straight forward, to flex and raise high first one knee and then
the other, and to continue stepping in the same position without too
much strain and at normal walking speed (about 110 steps per minute)
for a total of 50 or 100 steps (on occasion 200 or 300) (Fig. 8.1). The test
must be performed in a quiet room without lateral illumination, and the
examiner must not speak to the subject during stepping, in order to

avoid giving him any indication of possible changes in his position.

During the stepping test, the following items should be observed: a) presence or absence of swaying and its direction if present; b) changes in relative position of the head and body; c) deviation of both arms from the original position (stretched forward) in the upper, lower, or lateral direction; and d) locus of movements of both feet.

When the prescribed number of steps has been completed, the subject is asked to stop stepping and to stand upright in the last position. Then the angle of rotation of the body around its vertical axis and the distance as well as the direction of displacement of the body from the original position, if they are present, are measured by means of the circles and lines drawn on the floor (Fig. 8.2). When rotation or displacement of the body occurs during stepping, the phenomenon is called stepping deviation, and its parameters are expressed by an angle of rotation, an angle of displacement, and a distance of displacement as shown in Figure 8.3.

8.2.2. Results of the Stepping Test in Normal Subjects

According to Unterberger, normal persons show hardly any rotation of the body during *Tretversuch* unless labyrinthine stimulation is applied, but when the stepping is continued for several minutes, normal right-handed persons may show slight left rotation of the body, and normal left-handed ones slight right deviation of the body.

From my own experiences with the stepping test, it can be said that normal persons with intact hearing usually show no marked rotation of the body during stepping. At most they may show only slight forward progression and slight rotation of the body. Results obtained by examination of 500 normal persons who took 50 or 100 steps are summarized as follows:

1) Most normal subjects could complete 50 or 100 steps in the original position of the body.

2) Forward progression of the body up to 50 cm and 1.0 m after 50 steps and 100 steps, respectively, could be observed in some normal persons. An angle of rotation within 30° on each side and 45° on each side at the end of 50 steps and 100 steps, respectively, could be observed as well.

3) Backward displacement of the body during stepping was rarely observed in normal persons.

8.3. Stepping Deviation in Cases of Labyrinthine Stimulation

If the stepping test is performed in the period when eye nystagmus is present after rotatory or caloric stimulation of the labyrinth, displacement and rotation around the vertical axis of the body in the direction of the slow phase of nystagmus can be observed. Unterberger and Hirsch made the same observation on stimulation of an ear with 10 cc of water at 27°C or 5 cc of water at 15°C. Worthy of note in their description is that *"Körperdrehrektion"* or "spinning or waltzing" could be observed prior to the appearance of nystagmus and outlasted the duration of nystagmus. They believed that this was objective evidence of latent nystagmus or *Nystagmusbereitschaft*.

In Chapter 5, I reported the fact that deviation in vertical writing with eyes covered was induced by such a mild rotatory stimulation as incapable of inducing nystagmus and that the direction of deviation in writing in that case was often opposite to that seen in the period of manifest nystagmus, the latter being in accord with the slow phase of nystagmus when this was induced by stronger rotatory stimulation in the same direction.

Similar facts to those revealed by the writing tests were observed in the stepping test. Representative cases will be presented here, in which the test was performed after applying rotatory stimulation to the labyrinth as well as after changing the pressure in the external auditory meatus.

8.3.1. Rotatory Stimulation

Two turns in 10 seconds: A mild degree of rotation of two turns in 10 seconds ($2 \times /10''$) is given to a normal subject with intact hearing and eyes closed. Immediately on stopping rotation he is asked to open his eyes, to alight from the chair, to stand upright closing his eyes again, and to begin stepping according to the routine method. As stepping goes on, displacement and rotation around the vertical axis of the body in the direction of rotation of the chair take place. The rotation of $2 \times /10''$ is so mild that no postrotatory nystagmus or, at most, minimal postrotatory nystagmus with a low frequency and a short duration can be induced. Therefore, if nystagmus is adopted as the sole indicator of the labyrinthine reaction as in classical thinking, the labyrinth may hardly be considered to be stimulated by such a mild rotation. But this assumption is not true, since functioning of the vestibulospinal pathway is objectively evidenced by the fact that the stepping deviation can be observed for 2 or 3 minutes after cessation of rotation when postrotatory nystagmus was not present or when slight postrotatory nystagmus once present had disappeared. It should be noted that the direction of rotation of the body is the same as that of chair rotation.

Ten turns in 20 seconds: Immediately after a rotation of ten turns in 20 seconds ($10 \times /20''$), a subject is asked to alight from the chair and to step with his eyes closed. When the rotation is $2 \times /10''$, the stepping is carried out very smoothly and no ataxic tendency is observed, whereas when the rotation is $10 \times /20''$, the stepping is quite ataxic and the subject progresses staggering in the opposite direction to chair rotation, the locus of the feet showing zigzag lines. Rotation of the body in the opposite direction to chair rotation is also observed. Such deviation of the body is not a new fact, but has been familiar to us in performing the Romberg test, the gait test, or the pointing test. The ataxia and the deviation of the body in stepping disappear almost simultaneously with cessation of postrotatory nystagmus, and the subject becomes able to step in an almost normal way. However, on further continuation of the stepping, a new deviation of the body often occurs. This deviation is in the direction opposite to the one seen immediately after rotation; i.e., the second deviation—displacement and rotation of the body—is in the direction of chair rotation. Thus this deviation has the same characteristics as the stepping deviation following the rotation of $2 \times /10''$. The fact that two distinct kinds, or phases, of the reflex can be elicited according to variation in intensity of labyrinthine stimulation has been described in the previous chapters on vertical writing with eyes covered. The same fact has been reaffirmed here by means of the stepping test. One of the two phases of the reflex is observed when ataxic displacement and rotation of the body in the opposite direction to chair rotation (i.e., the direction of the slow phase of postrotatory nystagmus) occur con-

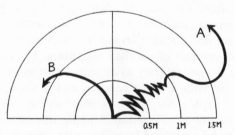

FIG. 8.4. Stepping deviation after rotation to the left (path of 50 steps). A, after 10 turns in 20 seconds; B ,after 2 turns in 10 seconds.

currently with manifestation of postrotatory nystagmus after rotation of
$10 \times /20''$. Another phase of the reflex, which is characterized by a direction of stepping deviation opposite to the former, is observed when displacement and rotation of the body in the direction of chair rotation
occur after rotation of $2 \times /10''$ as well as following cessation of postrotatory nystagmus induced by rotation of $10 \times /20''$.

It would be impossible to determine the presence of these two phases
if tests employing experimental eye nystagmus as the sole indicator were
used. The occurrence of spinal reaction unaccompanied by experimental
nystagmus, which is shown by the latter phase, is a fact of much importance from the physiological as well as pathological point of view.
A representative case in which the chair was rotated to the left is presented in Figure 8.4.

8.3.2. Changes in Pressure in the External Auditory Meatus

In the presence of a fistula in the bony wall of the labyrinth, nystagmus
can be induced by suction or positive pressure applied to the external
auditory meatus, and the directions of nystagmus with suction and
pressure are opposite to each other. This fact is well known under the
name of the fistula symptom. In normal persons with intact hearing, no
nystagmus can be induced by changes in pressure in the external auditory
meatus. Thus such a procedure has been considered incapable of yielding
labyrinthine stimulation in normal persons.

However, by performing the stepping test on normal persons with
intact ears, it has become clear that changes in pressure in one external
auditory meatus can stimulate the labyrinth and induce stepping deviation. As shown in Figure 8.5, when a pressure of $+100$ mmHg is

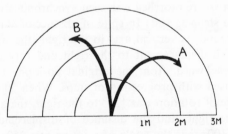

FIG. 8.5. Stepping deviation with changes in pressure in right
external auditory meatus (path of 120 steps). A, pressure
of $+100$ mmHg; B, pressure of -50 mmHg.

applied to one external auditory meatus, the stepping deviation (consisting of displacement and rotation around the vertical axis of the body) occurs toward the side of that ear, and when a pressure of -50 mmHg is applied to one external auditory meatus, the deviation is toward the side opposite that ear. During stepping, neither eye nystagmus nor ataxic attitude can be observed. Thus the stepping deviation continues to occur at a constant rate in parallel with increase in total steps up to 200, 300, or more, when the pressure in one external auditory meatus is changed to, and maintained at, a certain abnormal level.

8.4. Stepping Deviation in Patients Suffering from Ear Diseases

Reports have already been made by Unterberger and Hirsch on the spontaneous and experimental stepping deviation observable in cases with obvious signs of inner ear impairment. The fact to be reported here is a new one. The presence of imbalance of the vestibular function in cases of middle ear diseases without complications in the inner ear has been evidenced by the fact that pronounced spontaneous deviation in stepping is observed in such cases.

8.4.1. A Representative Case of Otitis Media Purulenta Acuta

A 60-year-old male visited our clinic complaining of right otalgia lasting 5 days and right otorrhea lasting 2 days. He also complained of impairment of hearing and tinnitus, but not of headache, vertigo, or ataxia.

Clinical findings: The osseous portion of the right external auditory meatus was reddened and swollen. The right drum was diffusely red with maximal redness in the superior quadrant and had a perforation of its anterior portion where otorrhea was seen synchronously with the pulse.

Results of the stepping test: In spite of absence of ataxia, the subject's body turned around its vertical axis to the right (the side of the lesion) concurrently with displacement to the right and forward, and then to the right and backward, from the original position. This deviation of the body increased with progression of steps. When 50 steps were completed, the angle of rotation was 120° to the right, the angle of displacement 30° to the right, and the distance of displacement 75 cm right forward. After 100 steps the angle of rotation was 350° to the right (almost one full turn to the right), the angle of displacement 105° to the right, and the distance of displacement 25 cm right backward.

8.4.2. A Representative Case of Obstruction of the Auditory Tube

A 37-year-old male visited our clinic. His left drum was found to be turbid and depressed. His right drum was quite normal. He had often been treated by catheterization when he felt a choking sensation and impairment of hearing of the left ear.

Results of the stepping test prior to catheterization: No ataxia could be found in stepping. While continuing stepping, however, his body showed gradual rotation around its vertical axis to the left (the side of the lesion) and displacement in the left forward direction. When 50 steps were taken the angle of rotation was 70° to the left, the angle of displacement 0°, and the distance of displacement 50 cm forward. When 100 steps were completed the angle of rotation was 180° to the left (i.e., exactly opposite to the original one), the angle of displacement 30° to the left, and the distance of displacement 100 cm left forward.

Results of the stepping test after catheterization: No ataxia was observed, as prior to catheterization. The marked left rotation of the body around its vertical axis, which had been observed before catheterization, was no longer demonstrated, and the angle of rotation was only 5° to the left when 100 steps were made. This angle is within normal limits of variation. Details were as follows: the angle of rotation 0°, the angle of displacement 0°, and the distance of displacement 50 cm forward at the end of 50 steps; and the angle of rotation 5° to the left, the angle of displacement 0°, and the distance of displacement 100 cm forward at the end of 100 steps.

8.4.3. Evaluation of the Two Cases

It should be noted that in each of these cases (prior to treatment) the rotation of the body around its vertical axis continued at a constant rate as stepping went on up to 200, 300, or more steps in total. This is essentially different from the results of experimental labyrinthine stimulation. When rotatory or caloric stimulation is applied to the labyrinth, the rotation of the body is striking at first, but it gradually becomes less marked until at last stepping is made normally without any rotation of the body. Also, when the labyrinth is stimulated by pressure, removal of pressure can restore normal stepping. It has not hitherto been considered that in middle ear diseases without complications in the inner ear, the labyrinthine function can be affected and its imbalance observed. The stepping test has revealed the presence of labyrinthine imbalance in such cases. The facts that the stepping deviation which can be observed prior to catheterization disappears following catheterization and that rotation of the body around its vertical axis can last as long as stepping continues obviously indicate the presence of labyrinthine imbalance in such cases.

8.5. Discussion

From the time of Bárány, special emphasis has been placed on observation of nystagmus in the examination of the labyrinthine function. Above all, experimental nystagmus induced by rotatory or caloric stimulation has played the most important role in the labyrinthine test. Also, to examine the vestibulospinal reflex (deiterospinal reflex), deviation on rotatory or caloric stimulation has been the main object of observation. In every report on the pointing test, *Abweichreaktion* and *Armtonusreaktion*, as well as in the original reports on the stepping test by Unterberger and Hirsch, the deviation induced by caloric stimulation of the labyrinth in clinical cases has been described. These reporters observed the fact that deviation was induced (i.e., the vestibulospinal reflex was brought into play) prior to appearance of experimental nystagmus and outlasted it, and they called the stage of manifest deviation unaccompanied by nystagmus "*Nystagmusbereitschaft*" or "latent nystagmus." However, observation of the deviation produced by rotatory or caloric stimulation, by which nystagmus can also be induced, seems necessarily to have drawbacks in the examination of labyrinthine function as compared with the observation of nystagmus, because the deviation cannot be measured so accurately as the duration and frequency of nystagmus. Experimental nystagmus shows a distinct beginning and ending, whereas the vestibulospinal deviation has no clear-cut beginning and ending. Though the deviation and ataxia cease to be marked at the time when nystagmus stops, the deviation continues for some time thereafter, gradually changing to the normal state without deviation. Thus, as far as clinical examination of the labyrinthine function by means of rotatory or caloric stimulation is concerned, no better indicator of the labyrinthine reflex is present than eye nystagmus; the vestibulospinal reflex expressed by deviation is less important.

It is the author's opinion that the test of vestibulospinal function becomes of great significance in clinical labyrinthine examination only when emphasis is placed on spontaneous deviation and not on the deviation induced by various stimulating procedures. The presence of imbalance of the labyrinthine function in cases of middle ear diseases without inner ear complications has been evidenced by striking spontaneous deviation seen in vertical writing with eyes covered (as stated previously) and in the stepping test (as described here). In these cases experimental nystagmus has shown almost no difference between the two sides. Thus spontaneous deviation is superior to experimental nystagmus in that the former can make manifest the latent imbalance of the labyrinthine function due to ear diseases, whereas the latter cannot. It has generally been considered that the labyrinthine reflex is brought into play first at the moment when rotatory or caloric stimulation induces experimental nystagmus. But such a consideration is not correct. The fact that deviation is seen in the vertical writing test and in the stepping test in cases of unilateral ear diseases indicates the presence of imbalance in tonus of the entire body musculature, which is caused by imbalance of the labyrinthine function due to unilateral ear diseases. As described above and in the previous communication, the deviation in vertical writing and in stepping seen in cases of obstruction of the auditory tube disappeared after catheterization. This disappearance of deviation does not mean cessation of functioning of the labyrinth, but it really means restoration of normal labyrinthine balance. Thus it is obvious from these facts that the vestibular labyrinth is always functioning whether experimental rotatory or caloric stimulation is applied or not, and that it is continuously regulating the tonus of the entire body musculature during performance of various voluntary movements such as writing vertically or stepping in the same position.

Another fact of physiological importance has been revealed by the stepping test (as well as by vertical writing with eyes covered). The stepping deviation can be evoked by such weak labyrinthine stimuli as are incapable of inducing experimental nystagmus, e.g. rotation of $2 \times / 10''$ and changes in pressure in the external auditory meatus (which cannot of course induce nystagmus). The stepping deviation unaccompanied by eye nystagmus has heretofore been referred to under the name of "*Nystagmusbereitschaft*" of "latent nystagmus." These names indicate less importance of this state than the state of manifest nystagmus. It is apparent that such a consideration is wrong in view of the fact that the direction is quite reverse between the stepping deviation induced by strong stimulation and accompanied by nystagmus, and that induced by weak stimulation and unaccompanied by nystagmus. The direction of deviations in stepping and in writing seen during manifestation of nystagmus is toward the slow component of nystagmus just as in cases of the Romberg test, pointing test, *Abweichreaktion*, etc. However, the direction of deviations in stepping and in writing induced by such mild labyrinthine stimulation as to be incapable of producing nystagmus is toward the rapid component of nystagmus which would be induced if stronger stimulation of the same modus were applied or if the fistula symptom were present. In this case (of mild stimulation) neither vertigo nor ataxia is present and the writing and stepping attitude is quite normal and yet deviations in stepping and in writing do occur.

This fact may be of great significance from the point of view of labyrinthine physiology. As described above, when positive pressure is applied to one external auditory meatus, the stepping deviation occurs toward that ear and when the pressure is negative, the deviation is opposite to that ear. If an analogy is permissible between a human subject and a flying bird, the rationality of this fact may well be supposed. When a bird is flying across the wind, the external auditory meatus facing the wind receives positive pressure and the other external auditory meatus receives negative pressure. Such changes in pressure in the external auditory meatus can induce changes in the tonus of the body musculature so as to elicit deviation of the body toward the wind, thus making straight flying possible although the wind is blowing from the side. This assumption may not be out of place from the evolutionary point of view. As to the stepping deviation, induced by a mild degree of rotation of $2 \times /10''$, which occurs in the direction opposite to the deviation immediately following the rotation of $10 \times /20''$, I am of the opinion that it has a certain unknown significance of physiological importance.

These two phases of the reflex, of which the presence was first clearly evidenced by the vertical writing test and the stepping test, have been entitled by the author the "stage of coordination" and the "stage of disturbance," respectively. The latter is the period in which experimental eye nystagmus, heretofore the most important phenomenon in the labyrinthine test, is manifest and the labyrinthine equilibrating function is disturbed, as indicated by transient severe ataxia apparent as a positive Romberg phenomenon. The former phase has hitherto been described vaguely under the name of latent nystagmus, but it is by the writing test and the stepping test that its presence has first been shown objectively. The labyrinthine function acting in daily postures as well as coordinated movements is considered to be acting in this phase of the reflex, which is observed in cases of mild rotation and changes in pressure in the external auditory meatus. Spontaneous deviation in stepping as well as in vertical writing seen in cases of ear diseases without inner ear complications may be a pathological condition seen in this phase of the reflex. Since my first report on the presence of these two phases of the reflex, its presence has been confirmed electromyographically in Japan in men and animals in which rotation or changes in pressure in the external auditory meatus are made.

The following is my opinion on labyrinthine physiology in general.

The essence of the labyrinthine physiology is that the labyrinthine reflex functions and elaborates the performance of action in the physiological process of voluntary movements in regulating the tonus of the entire body musculature unconsciously and ingeniously in response to various stimuli received by the body. Examination of experimental eye nystagmus is indeed an excellent method of testing labyrinthine function in clinical patients, and it often affords dramatic findings. It must be noted, however, that in examining nystagmus we are not observing the labyrinthine function in the completely physiological stage. During manifestation of experimental eye nystagmus, the subject can neither stand upright (Romberg's phenomenon), nor walk straight forward, nor gaze upon a resting object; he is obviously in a state of ataxia (stage of disturbance). In the nomal physiological state, the labyrinthine function is subordinate to and assists the will in regulating the entire body musculature, including muscles of the eyes and extremities. When experimental eye nystagmus is induced by intense labyrinthine stimulation, the labyrinthine reflex comes to overwhelm the power of the will and disrupts its normal regulation of the entire body musculature, thus producing ataxia. It should be borne in mind, therefore, that experimental eye nystagmus, despite its excellence in obtaining objective and reliable results, is observed only in a pathological state, though produced transiently, and not in the quite normal state viewed from the standpoint of labyrinthine physiology or physiology of body equilibrium.

Chapter 9

Recent Evaluation of Vestibular Function

9.1. Traditional Evaluation and Autonomic Nerve Reflex

In this section, I will introduce new trends in recent methods of evaluating vestibular function. I will then present some criticisms of these methods as well as my own method of evaluation.

Vestibular function has hitherto been evaluated by examining the reflexes provoked by labyrinthine stimulation. It has been confirmed by detailed work of pioneer researchers that the labyrinthine reflex affects the eye muscles, skeletal muscles throughout the body, and the smooth muscles of internal organs. This means that the labyrinthine reflex was classified into the following three reflexes, which were examined separately: the vestibulo-ocular reflex (eye muscle reflex), vestibulospinal reflex (skeletal muscle reflex), and vestibulo-autonomic nerve reflex (smooth muscle reflex). For instance, methods were developed for examinations of the labyrinthine reflex of the eyes, the extremities, and so forth. Examinations of nystagmi, such as postrotatory nystagmus and caloric nystagmus, test the eye reflex; whereas examinations with a goniometer and examination of walking, pointing, and stepping test the skeletal muscle reflex.

At present no tests of the autonomic nervous reflex are available, for although this reflex does exist, its manifestation and mechanism are not thought to be physiologically uniform. The same is true of other labyrinthine reflexes. Expression of the labyrinthine reflex, whether as an ocular reflex or a skeletal muscle reflex, is complex and depends on the mode and intensity of stimulation. In the past, only reflexes that are relatively uniformly manifested were examined. However, it must be remembered that these are only parts of the labyrinthine reflex, which has complex and seemingly contradictory expression depending on the intensity of stimulation. Although there is no problem in using certain labyrinthine reflexes as markers of vestibular function, it must be remembered that these markers do not represent the entire labyrinthine physiology. I believe that the true physiology of the labyrinth will be understood only by scientific observation and integration of labyrinthine reflexes induced by various stimuli, even though these reflexes may at first appear contradictory. The multiplicity of the labyrinthine reflex mentioned above is well represented by the autonomic nervous reflex.

As stated above, the labyrinthine reflex is mainly evaluated by examining the skeletal muscle reflex and ocular reflex, and especially nystagmus. Bárány (1907a) first evaluated nystagmus. Since then, various criticisms of Bárány's method have been raised and new tests, such as cupulometry, have been proposed. But even with new methods, it is nystagmus that is evaluated. The reason for this will be considered below.

In general, a reflex is understood to be an involuntary movement; voluntary movements and reflex movements are clearly distinguished in physiology. Nystagmus is, in this sense, a pure reflex movement. Nystagmus provoked by rotation or temperature is a reflex movement of the eyeballs produced by labyrinthine stimulation, and it cannot be suppressed voluntarily. Therefore, it is justifiable to assume that pure labyrinthine reflexes causing nystagmus take place in the muscles, and thus that pure labyrinthine function, unaffected by cortical or psychosensorial components, can be evaluated best by tests of nystagmus. In contrast, it is difficult to observe pure reflexes of the skeletal muscles of the limbs. In goniometry, for instance, the equilibrium reflex of the labyrinth that maintains the standing posture in spite of inclination is evaluated. While standing, skeletal muscles, particularly those in the legs and back, are tensed by the will to stand, i.e., a cortical component. A healthy normal person can remain standing on a considerable inclination, such as 25–30 degrees, but a person whose labyrinthine function is lost will fall even at an inclination of 2–3 degrees. The skeletal muscles of a normal person are tensed by volition to stand, and their tension is further adjusted and controlled by labyrinthine stimulation, resulting from the inclination. In other words, the labyrinthine reflex aids volitional movements in response to stimulation; it does not act independently on muscles, but in cooperation with volition by adjusting and aiding volitional movements.

Unlike in the classic concept of labyrinthine function, labyrinthine reflexes affect volitional movements. It is obvious from the fact that a person whose labyrinthine function is abolished falls down even at an angle of 2–3 degrees that the labyrinthine reflex works in a healthy normal person to maintain the standing posture.

In various tests of deviation, used to evaluate laterality or imbalance in labyrinthine function, certain voluntary movements are performed, such as walking, stepping, finger-pointing, and writing with eyes covered, although no so-called labyrinthine stimulation, such as rotation, is given. Involuntary deviation during these movements indicates imbalance in labyrinthine function. In this way, labyrinthine function is evaluated with respect to the skeletal muscles of the arms and legs, which do not move as involuntarily as eye muscles when complemented by the labyrinthine reflex.

Nystagmus provoked by labyrinthine stimulation is very important and useful in evaluation of labyrinthine function. This method of nystagmus evaluation, introduced by Bárány, has frequently been used as a diagnostic tool. It has also been used in research as means of studying labyrinthine physiology; nystagmus is produced by such methods as rotating a subject ten times in 20 seconds (postrotatory nystagmus), or pouring cold or warm water onto the tympanic membrane. Various other methods have recently been devised, but their basic principle is essentially that of Bárány's method. In recent reports, nystagmus and related phenomena have been considered as pure expressions of labyrinthine function. Particularly noteworthy is a trend to emphasize the sense of rotation accompanying nystagmus. This is examined by cupulometry, in which the continuation of nystagmus is measured, and the labyrinth is treated as a sensory organ. This trend is evident in the research of Grahe (1927) and in the recent development of audiology.

My views on recent methods for evaluation of labyrinthine function by rotatory and caloric stimulation are described below.

9.2. Rotatory Stimulation

Steinhausen (1925) showed that the cupula was deformed in the hecht (a kind of shark). Since that time, the main stream of studies on labyrinthine physiology has been dominated by the lymph flow theory, and various tests have been devised based on this theory. Bárány devised a method to evaluate postrotatory nystagmus. Recently, his method has been reassessed. In Bárány's method, positive angular acceleration is given initially, and this, according to Buys and Ryland (1939), will exert effects for 120 seconds. Buys claims that it is necessary to subtract the reflex caused by this initial acceleration from the postrotatory nystagmus provoked by the 20-second stimulation. On the basis of this reasoning, a subject is rotated first at a slow, subliminal angular acceleration velocity, and the velocity is then gradually increased. The object of this procedure is to avoid per-rotatory nystagmus caused by the initial angular acceleration in Bárány's method and observed during the rotation. The subliminal

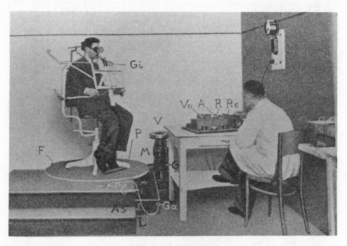

FIG. 9.1. Tönnies chair.

rate of acceleration has not yet been determined, but is supposed to be in the neighborhood of 1 degree/sec^2, and not more than 6 degrees/sec^2. Different types of apparatus have been devised to reduce the initial angular acceleration as much as possible. Examples of such types of apparatus are the cupulometer, described later, and the Tönnies chair, shown in Figure 9.1, in which acceleration of 0.05 degree/sec^2 (Buys, 1939), 0.3 degree/sec^2 (van Egmond, 1952 and Jongkees, 1953), and 0.5 degree/sec^2 (Fischer, 1930) are claimed to be possible. Starting with this acceleration, rotation is slowly increased to a constant velocity, which varies according to the worker: some use 180 degrees/sec (Fischer, 1930 and Arslan, 1953) like Bárány, whereas others use slower velocities, such as 20, 40, and 60 degrees/sec (Mittermaier, 1954). In cupulometry, even slower velocities of 2.5, 4, 6, and 18 degrees/sec are employed, and Jongkees (1953) reported that 60 degrees/sec is the strongest stimulus possible in the method. This constant rotation is stopped suddenly to evaluate postrotatory nystagmus, and in cupulometry to evaluate the after-sensation, or sensation of rotation. This procedure for rotation avoids the complication of the per-rotatory nystagmus, caused by a sudden initial, angular acceleration, during measurement of postrotatory nystagmus. During rotation, the vestibular organ and vestibular nervous tract are physiologically quiet, and the postrotatory nystagmus provoked in this way is a pure reflex and is thought to show the following characteristics: 1) regularity, 2) little individual variation, 3) a phasic rhythm, and 4) distinct slow phases.

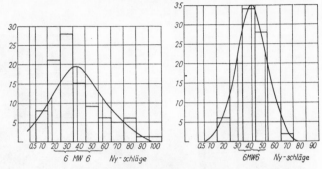

FIG. 9.2.   Comparison of results obtained by Bárány's method (left)
and by modified "rotacceleratory" stimulation (right).
Note the statistically smaller variation and standard fluctuation of duration of postrotatory nystagmus in the right
figure than in the left. Reproduced from Buys and Ryland
(1939).

As shown above, Bárány's method of rotation was criticized, and sudden angular acceleration was replaced by subliminal acceleration followed
by a constant velocity rotation. Jongkees (1953) propose the name "rotacceleratory" stimulation rather than "rotatory" stimulation for this.
These criticisms are all based on the theory of cupula deformation, and
they also have shortcomings. The criticisms are based on the idea that the
cupula is damaged by Bárány's rotation, because cupulometric values
are markedly reduced for several hours to several days after Bárány's
type of rotation. This idea is far-fetched and has not been scientifically
proven. However, I think that cupulometry is a good method since it
gives statistically smaller variation and standard fluctuation and less
frequently causes troublesome postrotatory autonomic nerve reflexes,
such as nausea, paleness, and vomiting, than Bárány's method, according
to the findings of Buys and Ryland (1939) shown in Figure 9.2. However,
the Tönnies chair and cupulometry require large pieces of apparatus
that are not easily installed in general clinical laboratories at present.
Arslan (1953) recommends the following procedure as a substitute:
start rotating a subject very slowly, increase the speed of rotation gradually
until the angular velocity is 180 degrees/sec (as in Bárány's method),
maintain this constant velocity for two to three minutes and then stop
rotation suddenly, and observe postrotatory nystagmus. I think that this
method is acceptable, and recommend that it should be used so that all
workers can examine nystagmus under the same conditions of angular acceleration and velocity.

9.3. Caloric Stimulation

New methods have been developed in the field of caloric stimulation. The time between cold stimulation and the onset of nystagmus is termed the latent period and has been considered important. Recently, however, Fischer (1930), Veits and Kozel (1930), and Arslan (1953) argued that its measurement has no diagnostic importance and is not significant, because the latent period is determined, not by the labyrinth, but by the size of the antral bridge, the amount of air, and the speed of blood flow. Therefore, these workers propose that nystagmus should be examined from the moment when endolymphatic convection is supposed to take place. They do not measure the latent period and use the Veits method (Veits and Kozel, 1930), which is performed as follows (Figs. 9.3 a and b).

1) The head of the subject is flexed forward 30 degrees, thereby placing the horizontal semicircular canal correctly in the horizontal position. This position guarantees that no convection occurs in the endolymph because it is horizontal when caloric stimulation is given. A correctly calibrated goniometer is placed in the mouth of the subject so that the angle of the head can be adjusted exactly.

2) Ten ml of water at 20°C (or 10 ml of water at 47°C) is introduced into the posteriocranial area of the tympanic membrane by syringe with a 5 cm needle. This forward flexion position is maintained for one minute after injecting the water, and during this time the subject wears Frenzel's glasses with +20 D lenses.

3) One minute later, the subject's head is bent 90 degrees backwards for two seconds, and then observation of nystagmus is started; the duration, number, and frequency of nystagmus are recorded.

Arslan considers differences in the latent period to be due to individual differences in heat conduction and argues that after one minute heat change is conducted to the endolymph regardless of individual differences; thus measurement of the latent period is unnecessary. Ino developed a similar method called the z-test. He distinguished three components of the latent period: x, y, and z. Component x is the time required for the temperature change in the ampulla of the horizontal semicircular canal (bony labyrinth) to reach the membranous labyrinth via the exolymphatic space. Component y is the time required for the heat to travel from the membranous labyrinth to cause flow in the endolymph. Ino claimed that x and y together are equal to one minute. Component z is the time from the beginning of endolymphatic flow to the onset of nystagmus. Ino used Veits' method of cold stimulation, and

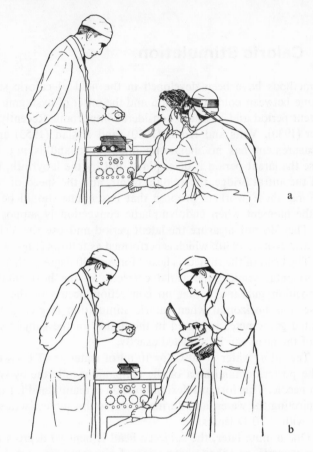

a

b

FIG. 9.3.   Veits' method for caloric stimulation of labyrinth and
observation of nystagmus. Reproduced from Veits and
Kozel (1930).

after this tilted the head backwards for eighty seconds and measured time
z, from head tilting to the onset of nystagmus. He found that in normal
persons it was 9–13 seconds; and that the duration of nystagmus was 45–
65 seconds. He claimed that the site of vestibular nerve impairment
could be determined from the time z.

9.4. Classification of Vestibular Function Tests

I have summarized above the new ideas and methods of examining nystagmus with rotatory and caloric stimulations. Next I will summarize

the trends in studies on labyrinthine physiology deduced by reviewing the literature on this subject.

First, the nature of stimulation should be considered. For example, slow rotation has been adopted on the basis of the idea that 10 rotations in 20 seconds is an unphysiologic, extreme stimulation. Since Steinhausen (1925) showed that the cupula was distorted, the angular acceleration is first low, and gradually increased to a high constant velocity, and then rotation is abruptly stopped for observation of nystagmus. In this method, labyrinthine stimulation is below the threshold both initially and during rotation; i.e., the labyrinth is not stimulated. When the high constant rotation is suddenly stopped, the cupula is distorted by inertia, and the labyrinth is stimulated for the first time, giving rise to nystagmus and the sense of rotation. The latent period is also analyzed in caloric stimulation (the term thermic stimulation was proposed by Portman and Cambrelin), and its validity has been claimed by some researchers.

I think these improvements in methodology were prompted by neurological considerations. In neurology, the vestibular system is comprised of the following four components:

1) Peripheral labyrinth or vestibular apparatus;
2) Vestibular root;
3) Vestibular nuclei;
4) Vestibular system above the vestibular nuclei or subtentorial vestibular pathways.

Neurologists try to characterize vestibular disturbance in terms of these components, and it was attempts to examine the function of each of them that resulted in improvements in the method. New methods have yielded data specific to these four components, and diagnostic approaches have been improved. However, it is still not yet possible to determine the exact locations of disturbances from the results of the new examination techniques; it is not possible to determine whether the vestibular nuclei or subtentorial pathways are impaired on the basis of these tests alone. Some otologists have confidently claimed that they have been able to do so, but in general the results of examinations of vestibular function are not sufficiently reliable for diagnosis of the location of impairment.

I have made critical comments on cupulometry in Chapter 14 of this book, but I will include a brief comment on the method here. After rotation, the chair does not revolve, but the sensation of rotation in the opposite direction is felt transiently. The chair is rotated at various velocities and the duration of this sensation is measured and tabulated: this record is called a sensation cupulogram. Simultaneously, the duration

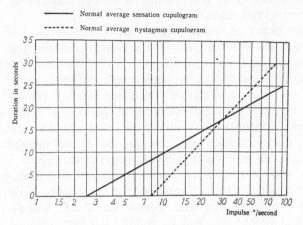

FIG. 9.4.   Sensation cuplogram and nystagmus cuplogram. Reproduced from Montandon *et al.* (1955).

of nystagmus is also measured, and is termed a nystagmus cupulogram. The two records are shown in Figure 9.4. The sensation of rotation after rotation has stopped is considered important and is treated as a labyrinthine sensation. This sensation is measured in the same way as an audiographic response. This after-sensation is provoked by physiological stimulation of the vestibulum and is considered to be a physiological and psychological reflex. From the viewpoint of the cupula distortion theory, this sensation may be considered physiological, but, in fact, is the sensation of simple vertigo and an illusion so far as common sense of physiology and psychology is concerned. Labyrinthine stimulation produces this queer sensation. It also produces the strange sensations of starting and stopping in linear movements, such as in elevators, which give sensations of sinking and floating.

The reason why after-nystagmus and after-sensation, or what I call vertigo delusion, are produced not just by vigorous rotation as in Bárány's method but also by slow rotation such as the one in recent cupulometry seems to me to be that the rotation is passively administered. The word labyrinthine stimulation conjures up the picture of rotating or moving experimental animals up and down, or passive changes in position. In labyrinthine physiology, it should be remembered that in the life of mammals there is no such stimulation as the ground rotating or moving violently up and down. Particularly, rotation around the longitudinal axis is never experienced, even during a big earthquake. This kind of stimulation is limited to laboratories. I think that labyrinthine function

TABLE 2.5.   Vestibular Function Examination

---

1. Examination of deviation
    Spontaneous nystagmus
    Standing, walking, and stepping tests
    Finger-pointing and writing with eyes covered
2. Righting reflex
    Standing on one foot and goniometer (balance platform) tests
3. Nystagmus
    Postrotatory nystagmus
    Caloric nystagmus
    Compression nystagmus
    Electric nystagmus

---

should be studied with respect to active movements and reflexes produced thereby which adjust the muscular tonus. The main symptoms of animals whose labyrinths have been destroyed are impairments in active movements. In humans with labyrinthine impairment, the main symptoms are also staggering and other forms of disturbed walking.

Based on the above considerations, I regard after-nystagmus and the after-sensation as transient motor ataxia and illusion. I found that repetition of this stimulation drastically reduced nystagmus and the after-sensation to practically zero in animals and men. The same holds for linear motion such as that in elevators. I found that the strange sensation of sinking and floating that most people feel when an elevator starts and stops is not felt by elevator operators who have been exposed to this sort of motion for a long period.

The theories of cupulometry and the gradual acceleration method of rotatory examination certainly have good points, but there is little sense in paying much attention to measurement of the after-sensation. Kobrak (1950) and Schierbeeck (1953) pointed out that measurements of the after-sensation are uncertain and fluctuate considerably. I consider that equilibrium function is the essential labyrinthine function and that it can be examined by testing the righting reflex. Imbalance between right and left labyrinthine function results in deviation in movements, which can be evaluated by examination. Strong and unusual stimulation of the labyrinth produces marked functional imbalance that disturbs physical equilibrium and causes nystagmus, which are easily measurable. On the basis of this concept, vestibular function tests can be classified as shown in Table 2.5.

Recently in reviewing the literature, I found many reports on detailed evaluation of nystagmus caused by Bárány's rotatory and caloric stimulations. Various arbitrary methods of examination have been devised. For example, in *Archives of Otolaryngology* 62: 5 (1955), two caloric tests were recommended, the Halpike-Fitzgerald method (Gutner, Gould, and Hanley, 1955) and the microcaloric test (Harrison and Lincoln, 1955). Although there are many new methods, none of them is much better than Bárány's original method. In these methods, the audiological approach and the attitude of treating the labyrinth as a sense organ and of measuring the labyrinthine sense are mixed up, and this makes interpretation of results very complicated, In short, recently, emphasis has been laid on nystagmus to the exclusion of other functions. However, the essential labyrinthine function is the equilibrium reflex, and imbalance in the labyrinth results in deviation in movement. Since these two are undeniable facts, examination of the equilibrium reflex and deviation should receive as much attention as nystagmus.

Recently, the question of the recovery period of nystagmus or *Nystagmusbereitschaft* has again received attention. However, this problem cannot be solved by studies only on the ocular muscles. I believe that this problem is closely related to deviation in skeletal muscles throughout the body, and cannot be solved without considering the latter. In conclusion, I think that functional examination should be performed according to Table 2.5; and as regards nystagmus examination, Arslan's rotation and Veit's caloric stimulation tests should be adopted for the reasons given above.

Part 3

Nystagmus

Chapter 10

After-Nystagmus in Children Who Have Received Special Training

10.1. Introduction

About 70 years ago, Hoshino reported that the duration and frequency of after-nystagmus in rabbits decrease progressively when the animals are rotated every day on a rotating chair. This was the first report in the world that dealt with the issue of repeated rotation and reduction in after-nystagmus. Follow-up studies were subsequently performed in various countries. Griffith (1920), for example, rotated mice two or three times daily at a speed of 10 rotations in 20 seconds and reported that after-nystagmus gradually decreased and finally disappeared. Other follow-up studies were conducted by Maxwell, Brucke, and Reston (1922), and others.

In this way, it was confirmed that in animals, repeated rotations result in reduction of after-nystagmus, although this fact has not been demonstrated in human beings. But it is well known that unlike untrained persons, skaters, dancers, and acrobats who have become used to rotation through constant training suffer slight, if any, adverse postrotation effects of relatively vigorous rotation, such as transient vertigo or equilibrium impairment of the body (e.g., Romberg's phenomenon), paleness, nausea, or vomiting.

A stimulus of rotation to which the body is not accustomed disrupts equilibrium or transiently induces an after-response. This after-response, which is brought about by muscle movement, is after-nystagmus and consists of a sense of being rotated in the direction opposite to rotation, i.e., vertigo, and abnormal tension in the organs of the vegetative nervous system, which causes paleness, nausea, and vomiting. In other words, in the period after rotation, during which the after-sensation is present, bodily equilibrium function is transiently impaired due to rotatory stimulation. Thus, abnormal vestibular responses appear transiently after rotation, when the vestibular labyrinthine function is disrupted by sudden,

137

excessive rotatory stimulation, although the vestibular labyrinth normally responds to rotatory stimulation so as to maintain bodily equilibrium by changing and regulating the muscle tone of the body.

Until the invention of airplanes, human beings were confined to two-dimensional movement on the earth. But airplanes subject humans to fast and complicated linear or rotational movements in three dimensions, a new experience. During flight, abnormally large angular and linear accelerations and sudden changes in atmospheric pressure and other physical conditions are said to cause impairment of physical equilibrium, and even to result in many unfortunate accidents. Aircraft crew members must maintain good bodily equilibrium by quickly responding to movements in three dimensions without developing vertigo or ataxia. It is important, therefore, to train the crew appropriately so that they develop good equilibrium function. With this in mind, we have been testing a new kind of training on grade school pupils, because their bodies are rapidly growing.

The main purpose of this training is to develop a tolerance to various accelerations to which the body is subjected during flying. Our method of labyrinthine function training is based on the physiology of body equilibrium and consists of inducing various active and passive rotational movements around the three axes of the body as well as linear movements.

By repeated daily training, the students improved remarkably in only a few months in skill, agility, and flexibility of movements. At the beginning of the training period, after-responses were rather strong. But as training progressed, the after-sensation—i.e., vertigo, transient equilibrium impairment, paleness, cold sweat, nausea, vomiting, and other labyrinthine after-responses—decreased in duration or disappeared completely. This was objectively confirmed in terms of reduction in the duration and frequency of after-nystagmus.

In the following experiment, we demonstrated reduction of after-nystagmus by repeated training in rotation. Our method will probably be useful for physical examination of air-crews, and this type of training should improve flight adaptability.

10.2. Comparison of After-Nystagmus of Students with and without Rotation Training

10.2.1. Subjects and Examination Conditions

The subjects were 82 boys 12 to 14 years old attending Sakai City Minato Grammar School who had been specially trained for four months.

The controls were 143 boys of the same age in a grade school in another city who had not been trained at all. None of them had hearing impairment or spontaneous nystagmus. Otoscopic examination showed that all of them had normal tympanic membranes. No cases of stenosis of the Eustachian tube were detected by Eustachian catheterization. None of the boys complained of tinnitus or vertigo.

The Sakai boys were trained for four months by the rotation training method devised by Hoshino and Fukuda (Fig. 10.1). They were supervised by the physical education teachers and by Dr. Hoshino, a member of our department at the Kyoto Imperial University School of Medicine. The training consisted of the following exercises:

a) Rotation around the frontal axis: somersaulting with arms straight, somersaulting after a hand-stand, jumping down and somersaulting, somersaulting and jumping down, somersaulting after a push-up and hand-stand, successive somersaults, somersaulting in pairs, hand-stand-jump somersault, somersaulting with one arm, push-up and hand-stand, and hoop movement.

b) Rotation around the sagittal axis: sideways rotation and hoop movement.

c) Rotation around the long axis: maximum-speed automatic rotation, circumference rotation, and rotation in a rotation chair.

The conditions of examination were as follows. A Kubo rotation chair was employed, and the subject was placed well back in the chair. His head was bent 30 degrees forward from the vertical position, and his head and trunk were fixed to the chair with belts so that they did not move during the examination. During rotation, the subject kept his eyes closed, and as soon as rotation was finished, he was told to open them and look steadily at the finger tip of the examiner, which was held 60 cm away at the level of the subject's eyes but 30 degrees from the median line in the direction of nystagmus.

The students were all examined in an auditorium (60×30 meters). The rotating chair was placed about 2 meters from a south window so that the light shone on the side of the subject's face. In this way, light was reflected at the sulcus sclerae, which made observation of nystagmus easier.

Three examiners who had had several years' experience with after-nystagmus observed nystagmus, and each examiner held two stop-watches. Special care was taken that the last nystagmus was not overlooked. An assistant timer reported the time every five seconds for 20 seconds, and one of the examiners rotated the chair smoothly at a constant speed of 10 rotations in 20 seconds, paying attention to the duration of rotation. When all three of the examiners recorded the same result, this result

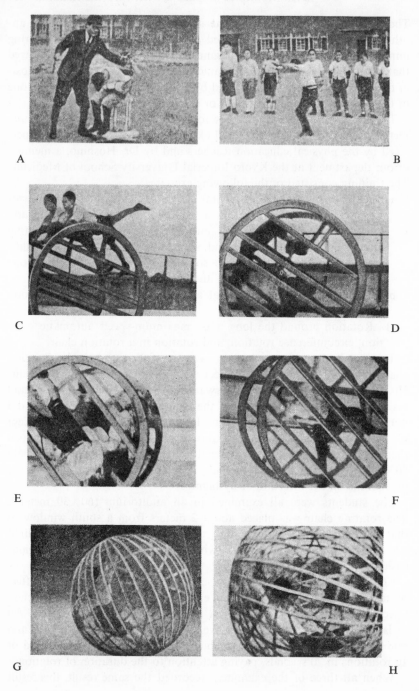

A

B

C

D

E

F

G

H

was taken. When two of the three were the same, this value was taken. When all three results were different, the arithmetic mean of the two most similar values was adopted. When the three values were equally scattered, the median value was adopted.

Rotation was first to the right and then, after four or five minutes of rest without change in the position of the head and body, to the left.

10.2.2. Results

Duration of after-nystagmus: After-nystagmus was slight, i.e., its duration was 15 seconds or less, in 30 (36.6%) trained students after rotation to the right and 22 (26.8%) after rotation to the left, whereas in untrained subjects the corresponding numbers were 6 (4.2%) and 8 (5.6%), respectively; thus the numbers of untrained students with slight after-nystagmus were far smaller. The average durations of after-nystagmus after rotation

---

FIG. 10.1.   Still photos from a 16-mm motion picture of rotation training.
A: Passive rotation exercises. A subject is rotated 5 times in 10 seconds or 10 times in 20 seconds with his head and trunk bent forward. Then he gets off the chair and stands up. When not accustomed to rotation, the subject develops marked equilibrium ataxia and vertigo. Not only is he unable to stand up, but he frequently slides off the chair with his head, trunk, and four limbs extended as soon as the rotation stops. As training progresses, he becomes able to put up with this stimulation and to stand up after it. This method, using Coriolis acceleration, was once employed by the Navy for fitness tests.
B: Active rotation exercises. A subject holds his arms up, bends his head forward, and revolves to the left or right as fast as he can. Then he raises his head and stands up. In the early days of training, he develops marked equilibrium ataxia because of Coriolis force.
C–F: Various exercises with a hoop or rolling frame devised at Minato Grammar School, Sakai, which enables four or five pupils to practice rotation exercises at the same time.
G and H: Exercises with a spherical rolling frame or hoop. In this frame, movement is in three, not two, dimensions, and can be passive as well as active. The frame seems to give excessive stimuli to the labyrinth; very few of the boys tested could perform the exercise the first time without developing equilibrium ataxia. This is a very sophisticated exercise.

TABLE 3.1. Duration of After-Nystagmus in Subjects with and without Training

Duration of after-nystagmus (sec.)	Rightward rotation		Leftward rotation	
	With training No. of cases (%)	Without training No. of cases (%)	With training No. of cases (%)	Without training No. of cases (%)
6–10	9 (11.0)	1 (0.7)	7 (8.5)	2 (1.4)
11–15	21 (25.6)	5 (3.5)	15 (18.3)	6 (4.2)
16–20	27 (32.9)	25 (17.5)	27 (32.9)	19 (13.3)
21–25	15 (18.3)	45 (31.5)	24 (29.3)	41 (28.7)
26–30	8 (9.8)	34 (23.8)	4 (4.9)	38 (26.6)
31–35	2 (2.4)	24 (16.8)	5 (6.1)	26 (18.2)
36–40		8 (5.6)		8 (5.6)
41–45		1 (0.7)		3 (2.1)
Total	82	143	82	143
Average (sec.)	17.88±0.67	24.55±0.54	18.79±0.64	25.13±0.57

TABLE 3.2. Eye Movements of After-Nystagmus in Subjects
with and without Training

Number of movements	Rightward rotation		Leftward rotation	
	With training No. of cases (%)	Without training No. of cases (%)	With training No. of cases (%)	Without training No. of cases (%)
1–10	8 (9.8)	7 (4.9)	4 (4.9)	6 (4.2)
11–20	18 (22.0)	13 (9.1)	19 (23.2)	17 (11.9)
21–30	34 (41.5)	27 (18.9)	26 (31.7)	31 (21.7)
31–40	15 (18.3)	31 (21.7)	23 (28.0)	26 (18.2)
41–50	5 (6.1)	21 (14.7)	6 (7.3)	26 (18.2)
51–60	2 (2.4)	25 (17.5)	2 (2.4)	20 (14.0)
61–70		12 (8.4)	2 (2.4)	11 (7.7)
71–80		2 (1.4)		3 (2.1)
81–90		4 (2.8)		2 (1.4)
91–100		1 (0.7)		1 (0.7)
Total	82	143	82	143
Average	24.64±1.24	37.70±1.42	27.68±1.37	38.07±1.42

to the right were 17.88±0.67 seconds in trained students and 24.55±0.54 seconds in untrained ones, and after rotation to the left, 18.79±0.64 seconds and 25.13±0.57 seconds, respectively. The difference between the averages for the two groups was 6.67 seconds for rightward rotation and 6.34 seconds for leftward rotation. These differences are statistically significant (Table 3.1).

Frequency of after-nystagmus: The frequency of after-nystagmus was relatively small, i.e., 20 times or less, in 26 (31.7%) trained students on rightward rotation and 23 (28.0%) on leftward rotation, whereas the corresponding numbers for untrained subjects were 20 (14.0%) and 23

(16.1%), respectively. The average frequency for the trained students was 24.64 ± 1.24 times on rightward rotation and 27.68 ± 1.37 times on leftward rotation, whereas the numbers for untrained students were 37.70 ± 1.42 times and 38.07 ± 1.42 times, respectively. The difference between the numbers for trained and untrained students was 13.06 times on rightward rotation and 10.39 times on leftward rotation. Thus the number was significantly smaller in trained students (Table 3.2).

10.3. Change in After-Nystagmus as a Result of Special Training

10.3.1. Subjects and Conditions of Examination

Thirty-six fifth-graders (12 to 13 years old) at Minato Grammar School, Sakai City, who had received special training were examined. All of them had normal hearing and a normal tympanic membrane on otoscopic examination. No stenosis was found on Eustachian catheterization, and none of them had tinnitus, vertigo, or spontaneous nystagmus. They received training for six months and were examined every two months. The method of training and the conditions of examination were the same as in the previous experiment.

10.3.2. Results

Duration of after-nystagmus: After-nystagmus was slight, i.e., its duration was 15 seconds or less, on rightward rotation in 3 students (8.3%) on the first examination on October 26, 1943; in 13 (36.1%) on the second examination on December 26, 1943; and in 15 (41.7%) on the third examination on March 3, 1944; after leftward rotation it was low in 4 (11.1%) on October 26, 1943; in 9 (25.0%) on December 26, 1943; and in 18 (50.0%) on March 3, 1944. Thus the number of students with low values increased as training progressed.

The average duration of after-nystagmus after rightward rotation was 20.97 ± 0.87 seconds on October 26, 1943; 17.09 ± 0.94 seconds on December 26, 1943; and 15.84 ± 0.81 seconds on March 3, 1944: after leftward rotation it was 21.25 ± 0.79 seconds, 17.91 ± 0.87 seconds, and 15.00 ± 0.71 seconds, respectively, at these times. These results show that the duration of after-nystagmus decreased as training progressed (Table 3.3).

Frequency of after-nystagmus: The frequency of after-nystagmus was relatively small, i.e., 20 times or less, after rightward rotation in 9 students

TABLE 3.3. Relation between Duration of After-Nystagmus and Progress in Training

Duration of after-nystagmus (sec.)	Rightward rotation						Leftward rotation					
	Oct. 26, 1943		Dec. 26, 1943		Mar. 3, 1944		Oct. 26, 1943		Dec. 26, 1943		Mar. 3, 1944	
	No. of cases	%	No. of cases	%	No. of cases	%	No. of cases	%	No. of cases	%	No. of cases	%
6–10	1	2.8	3	8.3	4	11.1			2	5.6	3	8.3
11–15	2	5.6	10	27.8	11	30.6	4	11.1	7	19.4	15	41.7
16–20	13	36.1	15	41.7	16	44.4	9	25.0	17	47.2	14	38.9
21–25	13	36.1	4	11.1	3	8.3	16	44.4	7	19.4	3	8.3
26–30	5	13.9	3	8.3	2	5.6	6	16.7	2	5.6	1	2.8
31–35	2	5.6	1	2.8			1	2.8	1	2.8		
Total	36		36		36		36		36		36	
Average (sec.)	20.97±0.87		17.09±0.94		15.84±0.81		21.25±0.79		17.91±0.87		15.00±0.71	

TABLE 3.4. Relation between Eye Movements of After-Nystagmus and Progress in Training

Number of movements	Rightward rotation						Leftward rotation					
	Oct. 26, 1943		Dec. 26, 1943		Mar. 3, 1944		Oct. 26, 1943		Dec. 26, 1943		Mar. 3, 1944	
	No. of cases	%	No. of cases	%	No. of cases	%	No. of cases	%	No. of cases	%	No. of cases	%
1–10	1	2.8	1	2.8	2	5.6	1	2.8	1	2.8	2	5.6
11–20	8	22.2	10	27.8	15	41.7	6	16.7	12	33.3	17	47.2
21–30	14	38.9	16	44.4	11	30.6	16	44.4	17	47.2	14	38.9
31–40	6	16.7	6	16.7	7	19.4	8	22.2	4	11.1	3	8.3
41–50	3	8.3			1	2.8	3	8.3	1	2.8		
51–60	2	5.6	3	8.3			2	5.6				
61–70									1	2.8		
71–80	2	5.6										
Total	36		36		36		36		36		36	
Average	30.00±2.41		25.83±1.88		22.23±1.70		28.34±1.19		23.89±1.26		20.00±1.44	

(25.0%) on October 26, 1943; in 11 (30.6%) on December 26, 1943; and in 17 (47.2%) on March 3, 1944: after leftward rotation it was low in 7 (19.4%), 13 (36.1%), and 19 (52.8%), respectively, at these times. Thus the number of students with a low frequency of nystagmus increased as training progressed.

The average frequencies of nystagmus for rightward rotation were 30.00 ± 2.41 times on October 26, 1943; 25.83 ± 1.88 times on December 26, 1943; and 22.23 ± 1.70 times on March 3, 1944: after leftward rotation the durations were 28.34 ± 1.19 times, 23.89 ± 1.26 times, and 20.00 ± 1.44 times, respectively. Thus the average nystagmus frequency gradually decreased as rotation training progressed (Table 3.4).

10.4. Summary

1) Comparison of postrotatory nystagmus of 82 Japanese boys 12 to 14 years old who had received four months of rotation training with those of 143 boys of the same age who had not received training showed that the duration and frequency of after-nystagmus were significantly reduced by training.

2) Examination of 36 healthy Japanese boys 12 to 13 years old every two months during training in rotation for about six months showed that the duration and frequency of postrotatory nystagmus gradually decreased during training.

3) Other workers have reported that postrotatory nystagmus is reduced when animals are rotated repeatedly, and they have explained this as the result of becoming accustomed to rotation (Griffith, 1920; Maxwell, Brucke, and Reston, 1922; and others), fatigue (Fischer and Babcock, 1919; and others), adaptation (Doege, 1923a), or a neck reflex (de Kleijn and Versleegh, 1930).

4) Acclimatization to repeated rotatory stimulation is also known from daily experience. No systematic study of this acclimatization has yet been performed, but our experiments on repeated rotation of humans gave results similar to those of animals experiments.

Chapter 11

Sports and Nystagmus

11.1. Introduction

In this chapter I will describe typical nystagmus, which has rapid and slow phases, from the standpoint of equilibrium physiology. The purpose is to evaluate the relationship between the roles of the labyrinth and the visual sense in human equilibrium function. These two are related in the sense that nystagmus is brought about by both labyrinthine and optic stimulations. In other words, the labyrinth and the visual sense are concerned with the equilibrium function of the body via nystagmus.

Regardless of whether it is labyrinthine or optic, nystagmus is brought about when there is excessive stimulation to body equilibrium. For instance, when the body receives rare stimulation of angular acceleration, as when rotated on a rotation chair, nystagmus is brought about in different directions during and after rotation. When objects move in front of the eyes, as when looking outside a moving train, nystagmus is brought about. This ocular reflex or nystagmus is interpreted as maintaining bodily equilibrium by trying to keep moving objects still on the retina. What is important here is that on both occasions, nystagmus is provoked by a strong stimulation that is unusual for the body, e.g., rotations of the body and moving of objects on the retina.

11.2. Postrotatory Nystagmus

Bárány's test of nystagmus is a labyrinthine function test in which the nystagmus produced by 10 turns in 20 seconds is evaluated. I entertain some reservations about this frequently conducted test; next I will explain these reservations and give my opinions about them.

As is well known, after rotation at a speed of 10 turns in 20 seconds, normal persons are said to have after-nystagmus for 15 to 45 seconds,

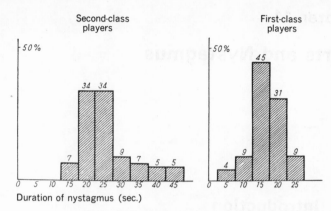

Duration of nystagmus (sec.)

FIG. 11.1. Duration of postrotatory nystagmus in first- and second-class ice hockey players after 10 turns in 20 seconds. Reproduced from Narita (1954).

and nystagmus of less than 15 seconds' duration is generally thought to indicate reduction of labyrinthine function. In fact, after-nystagmus lasts for less than 15 seconds and sometimes does not even develop in cases of loss of labyrinthine function and in some deaf-mutes. While the after-nystagmus is present, there is strong rotatory vertigo, the skeletal muscles of the body are ataxic, Romberg's phenomenon is present, and in extreme cases the person falls down. In other words, during after-nystagmus, equilibrium ataxia of the body is transiently present. But judging from the standpoint of equilibration—the essential labyrinthine function—it is not necessarily true that unless this ataxic condition lasts for 15 seconds or more, labyrinthine function is not normal or the organ of equilibration is dysfunctional.

With this consideration in mind, we examined various athletes who were thought to have particularly good equilibrium function. We found that the better the athlete, the shorter was the duration of after-nystagmus compared with that of non-athletes of the same age. In well-trained athletes, the duration was often less than 15 seconds. Narita (1954) compared first- and second-class ice hockey players at a certain university. The results in Figure 11.1 show that better athletes have a shorter period of after-nystagmus.

We trained fourth, fifth, and sixth graders at an elementary school in various rotation exercises of both passive and active kinds. After a few months, the duration of after-nystagmus in many children became less than 15 seconds, as shown in Figure 11.2. Associated with this change, their motor and equilibrium functions improved markedly. We took these

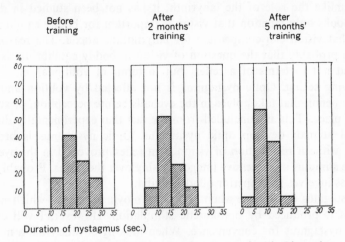

Duration of nystagmus (sec.)

FIG. 11.2.   Duration of postrotatory nystagmus in 4th–6th graders
(10 turns in 20 seconds). Left, before training; center,
after 2 months' training; right, after 5 months' training.

city-raised pupils to ski; although they had no prior experience of skiing,
they all became able to ski without falling within one day. Thus although
it is true that some persons in whom postrotatory nystagmus lasts for
less than 15 seconds do have reduced labyrinthine function, other persons
have excellent equilibrium function with an after-nystagmus of 15 seconds
or less. Bárány's test of postrotation nystagmus is a good clinical test, but
it should be borne in mind that during postrotatory nystagmus, bodily
equilibrium is impaired and vertigo is experienced. This is not a labyrinthine reflex within a physiological range of stimuli, but is a physiological aberration, so to speak. It must be realized that this reflex is produced by an excessive stimulus that is not normally experienced, e.g., 10
turns in 20 seconds. Our findings with athletes and school children show
that this fact must not be overlooked.

11.3. Optic Nystagmus

It is well known that vision is necessary for retention of bodily equilibrium. For example, a person with abnormal labyrinthine function
whose postrotatory nystagmus is less than 15 seconds can maintain equilibrium and perform exercises if his eyes are open; but if his eyes are closed,
thereby blocking vision, equilibrium is lost and motor ataxia develops.
The question of the role of vision in bodily equilibrium is thus important,

but, unlike the role of the labyrinth, it has not been studied in detail. Textbooks only mention that vision is important for bodily equilibrium, and that vision can compensate for labyrinthine ataxia. The reason for this is probably that the question of vision in bodily equilibrium is considered not in terms of a reflex, but in terms of volitional movement involving seeing. Optic nystagmus is not affected by volition and is a phenomenon that takes place in the eyeballs before perception, or seeing, takes place. This is concluded from the fact that experimentally decerebrated animals develop optic nystagmus. Optic nystagmus is brought about without recognition of vision by a reflex provoked in the eyeballs by the stimulus of successive images passing over the retina. Probably this process involves the superior quadrigemina.

Nystagmus takes place when observing moving objects—which means that it is produced by optokinetic stimulation. However, I will call it optic nystagmus for convenience. When looking outside from a moving train, nystagmus is produced in the direction of movement of the train. This is interpreted to be a reflex by which moving objects are kept fixed on the retina. This optic nystagmus is composed of a rapid phase in the direction of the moving train and a slow phase in the opposite direction. It is a typical nystagmus like the labyrinthine reflex in which the eyeballs move in a similar way with rapid and slow phases. In otological examination of labyrinthine function, the movement of the eyeballs produced by various labyrinthine stimuli, particularly nystagmus, has traditionally received attention. In some experiments, vision was blocked so that optic nystagmus did not interfere with labyrinthine nystagmus to determine the effect of labyrinthine function only on the eyeballs. However, change in body position or movements stimulates the labyrinth, and the labyrinth varies the tension of the eye muscles to cause nystagmus. Moreover, the labyrinthine reflex in turn moves the eyeballs so as to complete vision in spite of the change in the body position or movements so that bodily equilibrium is maintained. Therefore, in studies of equilibrium physiology, it is important to evaluate the relationship between the labyrinth and the visual sense as well as to study the labyrinthine reflex in the eyeballs apart from vision.

To provoke optic nystagmus easily, we used a drum-shaped cylinder. As shown in Figure 11.3, the cylinder was 75 cm in diameter and had eight red lines painted on it 30 cm apart. When the cylinder was turned horizontally, the subject looking at the cylinder developed horizontal nystagmus with its rapid phase in the direction opposite to that of rotation. When the cylinder was turned vertically, like a mill, vertical nystagmus with its rapid phase opposite to the direction of rotation was produced with the same rhythm; i.e., when the cylinder moved downward,

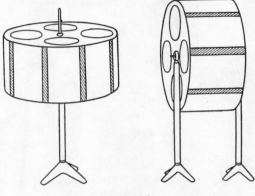

FIG. 11.3

the nystagmus was upward, and *vice versa*. We used this apparatus to study the mutual influence of labyrinthine nystagmus and optic nystagmus.

11.4. Influence of Optic Nystagmus on Labyrinthine Nystagmus

When a person is rotated on a rotation chair at a speed of 10 turns in 20 seconds, he develops horizontal nystagmus immediately after the rotation. The rapid phase of this nystagmus is in the direction opposite to that of rotation. The cylinder described above can be placed in various positions and rotated, and the examinee with labyrinthine nystagmus looks at it. The retina of each eyeball which is undergoing postrotatory nystagmus receives optic stimulation that can produce nystagmus. We studied the effects of simultaneous labyrinthine and optic stimulations on the eyes.

The cylinder was rotated upward, which should produce downward optic nystagmus, and an examinee with horizontal postrotatory nystagmus looked at it (Fig. 11.4). The resulting nystagmus was first horizontal, then gradually became diagonal and then downward, i.e., in the direction of the sum of the two vectors, and finally showed a rapid phase in the downward direction. This last nystagmus persisted for as long as the cylinder was rotated. When the cylinder was rotated in the opposite direction, similar changes ensued, except that in the vertical nystagmus the rapid phase was upward. Therefore, as expected, the direction of the

FIG. 11.4. A subject given optokinetic stimulation immediately after Bárány's rotation (10 turns in 20 seconds). The cylinder rotates at a constant speed of one rotation in 5 seconds, and thus the subject whose eyes are undergoing horizontal postrotatory nystagmus receives optokinetic stimulation which provokes vertical optic nystagmus.

resultant vector pointed diagonally and then upward.

Generally, in postrotatory nystagmus the frequency is high immediately after cessation of rotation and then gradually decreases to zero. On the other hand, the optic nystagmus produced when looking at the cylinder has a constant frequency and rhythm, because the speed of rotation is constant. Postrotatory nystagmus and optic nystagmus are caused by different stimuli and have different rhythms. And yet they are expressed as a combined vector force, the optic nystagmus finally taking over. What kind of anatomical and physiological mechanisms are responsible for this? This is a very interesting question to be solved.

Optokinetic stimulation is given immediately after Bárány's rotation. When rotation of the chair is stopped, postrotatory nystagmus develops immediately. The mill-shaped cylinder placed in front of the rotation chair rotates at a constant speed of one rotation in 5 seconds. This means that the eyes which are undergoing postrotatory nystagmus receive optokinetic stimulation which provokes vertical optic nystagmus.

The cylinder was then placed horizontally and rotated so that optic nystagmus would be in the same direction as postrotatory nystagmus. In this case, purely optic nystagmus and optic nystagmus combined with

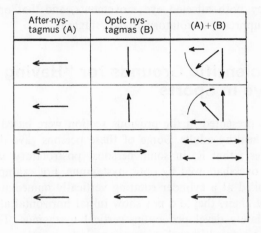

FIG. 11.5. Eye movements affected by postrotatory after-nystagmus
and optic nystagmus. Figures in right columns show tendency for optic nystagmus to suppress after-nystagmus.

postrotatory nystagmus were essentially the same. This means that the
optic nystagmus overpowered the postrotatory nystagmus from the
beginning.

Next, the cylinder was rotated so that the direction of the rapid phase
of optic nystagmus was opposite to that of postrotatory nystagmus. Immediately after rotation in a chair, the examinee looked at the cylinder.
The direction of the nystagmus thus produced was initially that of postrotatory nystagmus, but the direction then gradually changed to that of
optic nystagmus. The nystagmus immediately after rotation had longer
intervals and lower frequency than that produced by optic stimulation.
In the process of change to optic nystagmus, the eyeballs stopped moving
or moved like a pendulum—a noteworthy finding.

From these findings, we learned that labyrinthine after-nystagmus could
be affected by optic nystagmus and that the optic nystagmus could at
times suppress after-nystagmus. When the rotation of the cylinder was suddenly stopped in the middle of the examination, thereby eliminating optic
stimulation, the nystagmus resumed the initial direction of postrotatory
nystagmus. In other words, when the cylinder which was rotating vertically
was stopped, the eyeballs, which were moving vertically under the influence
of optic stimulation, resumed their original horizontal nystagmus. This
finding suggests the presence of suppression, as is indicated in Figure 11.5.
In this way, by studying the relationship between optic and labyrinthine
nystagmus, we learned that although the two were caused by different

stimuli, they both affected eye movements, and that optic nystagmus tended to suppress postrotatory after-nystagmus.

11.5. Scientific Grounds for "Having a Good Eye in Sports"

The results presented in the previous section were based on a study on about 100 healthy adults. Some of these persons gave different results from the rest; that is, in some persons, postrotatory nystagmus was completely overshadowed by optic nystagmus. For example, when these persons looked at a cylinder rotating vertically immediately after being rotated in a chair, they did not show initial horizontal after-nystagmus, but immediately developed optic, vertical nystagmus. Even when the cylinder was rotated in such a way as to induce optic nystagmus, in the opposite direction to after-nystagmus, these persons did not show any after-nystagmus, but promptly showed optic nystagmus with a rapid phase in the opposite direction. In other words, optic stimulation over-powered the after-nystagmus from the beginning. By overpowering, I mean that the postrotatory nystagmus did not develop unless the optic stimulation of looking at the rotating cylinder was not present, or rota-tion of this cylinder was stopped, i.e., optic stimulation was eliminated and the visual field became stationary; on stopping optic stimulation, the after-nystagmus in the direction of postrotatory nystagmus reap-peared. When these persons were studied carefully, it was found that many of them were sportsmen, particularly those who were good at ball games. From this we suspected that "having a good eye in sports" might mean that optic nystagmus suppressed postrotatory nystagmus. On the basis of this assumption, we examined various high school students who were good at sports, as well as famous baseball players from Japa-nese professional baseball teams. This examination showed that in these persons, postrotatory nystagmus was overshadowed immediately, or at least within five seconds. "Having a good eye in sports" is a vague concept based on experience, but we believe that it can be demonstrated objectively by showing that optic nystagmus overshadows postrotatory after-nystagmus.

11.6. Nystagmus and Vision during Rotation

When a person is rotated on a rotation chair with his eyes open, nystag-mus develops with its rapid phase directed towards the direction of ro-

tation. Even when his eyes are closed, nystagmus occurs in the direction
of rotation, and can be felt by touching the eyelids during rotation. The
nystagmus developing while the eyes are closed can be considered to be
purely labyrinthine and to be caused by the labyrinthine stimulation of
rotation. Therefore, the nystagmus that appears while the eyes are open
is caused by both optic and labyrinthine stimulation. A person with no
labyrinthine function does have optic nystagmus if the eyes are open,
as the objects projected on the retina move in one direction or there is
optokinetic stimulation. However, when his eyes are closed, the labyrinthine reflex, or the nystagmus during rotation, does not appear as
the labyrinth is afunctional.

Nystagmus during rotation is considered to be important in otological
science. In otology, nystagmus during rotation is considered to be the
essential labyrinthine reflex. But how do the eyeballs move when a person
is rotated on a chair with his eyes open while the objects around him
rotate at the same angular veolocity as he does so that their images remain stationary on his retina? Does he have rotational nystagmus caused
by labyrinthine stimulation because he is rotated even though his visual
field is stationary? To examine this problem, we devised the apparatus
shown in Figure 11.6. This is a large paper cylinder with a radius of one
meter that rotates like a merry-go-round. The cylinder is 1.7 meters
high and is covered with paper with 16 vertical black lines on it. A rotation chair is placed in the center of this cylinder, and the examinee sits
on the chair and fixes his eyes on one vertical line in front of him. The
chair and the cylinder rotate at the same angular velocity. Under these
conditions, the labyrinth receives the stimulation of rotation, but there
is no optokinetic stimulation because the cylinder rotates at the same
angular velocity as the chair.

What happens to the eyeballs during this experiment? The answer is
simple. The eyeballs remain stationary and there is no nystagmus, because, although the labyrinthine stimulation of rotation is present, nystagmus does not develop if there is no moving image on the retina.
In other words, nystagmus develops as a labyrinthine reflex during
rotation in order to keep images of things as stationary as possible.
When the visual field does not move, as in this experiment, nystagmus
does not develop. Another point I would like to raise is that in this experiment, vision overshadows the labyrinthine reflex of nystagmus. This
is proven by the facts that nystagmus develops when the eyes are closed
because of rotation, and that after the cessation of rotation of the chair
and cylinder, postrotatory nystagmus is always seen. From the standpoint of equilibrium physiology, the role of the labyrinth in relation to
the eyeballs is to support vision, and the labyrinth is subject to vision.

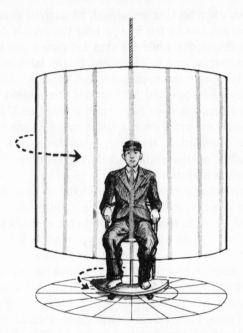

FIG. 11.6. Large rotating cylinder. For illustrative purposes, the cylinder is drawn as if it were transparent; actually, however, it is covered with opaque paper. One or two examiners stay inside the cylinder with the examinee. The chair and the cylinder rotate at the same angular velocity. If the examinee looks at a line in front of him, no nystagmus develops during rotation.

Nystagmus, particularly that during rotation, is the most important function in otology as an index of labyrinthine function. However, this nystagmus can be suppressed easily by change in optic stimulation, and this is important to remember in studies of equilibrium physiology and the mechanism of the reflex. Bodily equilibrium is the integral of various complicated reflexes, and it is important to pay attention to the relationship and combination of these individual functions, rather than to the individual reflexes, as shown by these experiments.

11.7. Optic Motor Ataxia and Vertigo

In the previous section, the cylinder was used only to study optic nystagmus. The examinee can assume various postures inside this rotating

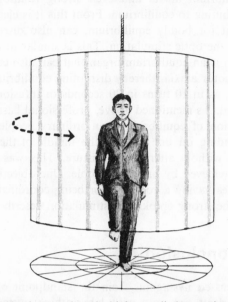

FIG. 11.7. Examinations conducted with large cylinder rotating at a constant speed of one rotation per 10 seconds. As illustrated, the examinee stands on one foot inside the cylinder or marks time. After a few seconds, motor ataxia and vertigo develop. Marked motor ataxia always develops in the direction of rotation. The labyrinth is not stimulated at all, and rotational vertigo and ataxia are of purely optokinetic origin.

cylinder, thereby provoking optic nystagmus as well as various skeletal muscle reflexes of optic origin. He may stand on both feet, or on one foot, or mark time in the center of this rotating cylinder, as shown in Figure 11.7.

Usually, after marking time a few steps, motor ataxia develops and the examinee staggers or falls. Marked vertigo also develops. This was observed at a high rate among healthy adults tested. Details of this finding will be presented in Chapter 12. Vision plays an important role in bodily equilibrium. Persons with little or no labyrinthine function can maintain equilibrium and perform movements just like normal persons, when their eyes are open. But when their eyes are closed, they lose balance. This shows that vision completely compensates for labyrinthine dysfunction. However, when the entire visual field receives images that appear successively, as when looking at a large rotating cylinder, vision brings about equilibrium ataxia and even strong rotationary vertigo, rather than contributing to equilibrium. From this it is clear that vision, which is important for bodily equilibrium, can also disrupt equilibrium, depending on the optic stimulation. This is similar to the case of the labyrinth, which is an equilibrium organ that can also cause vertigo, nystagmus, and motor ataxia, thereby disrupting equilibrium when stimulation is excessive, as in 10 turns in 20 seconds or infusion of cold water into the ear canal. As mentioned above, professional baseball players showed less disruption of equilibrium than ordinary people. One pitcher could remain standing on one foot in the middle of the cylinder for almost one minute without altering his posture. This was quite unusual, and was rarely achieved by ordinary people. One objective demonstration of good athletes having a good eye, or being coordinated, is that they can withstand the strong optokinetic stimulation described above.

11.8. Conclusion

I have discussed nystagmus from the standpoint of equilibrium physiology and dealt not only with labyrinthine nystagmus but also with optic nystagmus which has hitherto received little attention. I have also touched on the equilibrium-physiological relationship between the two. Although Bárány's postrotation nystagmus test is one of the best clinical labyrinthine tests, results of this test cannot be interpreted uniformly from the standpoint of equilibrium physiology. I have evaluated the significance of this test and studied first the relationship between postrotatory nystagmus and optic nystagmus and then between per-rotation nystagmus and optic stimulation. I showed that with excessive stimulation, vision, which normally contributes to bodily equilibrium, can disrupt and destroy equilibrium. I also demonstrated one of the objective factors for "having a good eye in sports."

Last, I would like to make reference to eye movements. When several outside objects move at a certain speed and are projected onto the retina, optic nystagmus develops, and it is termed an optic reflex. In watching a bird or an airplane flying high in the sky or when some other slow moving object is projected onto the retina, the eyeball slowly follows its movement. The eye moves effortlessly and unconsciously so as to keep the image in the middle of the retina. The mechanism of the rapid phase of nystagmus is still unsettled. Some consider it a reflex of the ocular muscles caused by contraction. By the same token, the slow movement of the eyeball without the rapid phase, as in tracking a moving object, can be considered a reflex, or nystagmus. One can think of this as a re-

FIG. 11.8. Simultaneous movement of eyeballs with head tilt to
retain original position.

flex or nystagmus without development of the rapid phase. When the
head is tilted to the side relative to the trunk, the eyeballs move slowly
simultaneously with the tilting and move in the opposite direction to the
head tilt so as to retain their original position (Fig. 11.8). This is a labyrinthine reflex caused by displacement of the otolith and is called *Gegenrollung* or counter-rolling. By analogy to this reflex, the movement of the
eyeballs in following a slowly moving object on the retina should be
called an optic ocular reflex, although it does not have a rapid phase.
The pyramidal motor area of the cerebrum, strangely, does not include
an area concerned with eye movements. Instead, there is an area inside
the extrapyramidal region that controls eye movements, which is called
adversive Felder. This suggests that the eye movement is essentially a
reflex movement caused by stimulation on the retina by one or more
moving objects.

Chapter 12

Optic Nystagmus and Posture

12.1. Optic Nystagmus

Numerous studies have been done on optic nystagmus, starting with those of Helmholtz and continuing with those of Bárány (1907b), Fischer and Babcock (1919), Weizsäcker, Hoshino and others. Rhythmic eye movements are brought about by looking at objects moving in front of the eyes, and are caused by the stimulation of these moving objects successively projected on the retina. This nystagmus was termed *Eisenbahn Nystagmus* (or railroad nystagmus) by Bárány, *optischer Nystagmus* or *optischer Drehnystagmus* by Ohm (1924), and *optokinetischer Nystagmus* by Fischer. This nystagmus is seen in daily life in the eyes of a person looking out from the window of a moving train, and so was called railroad nystagmus by Bárány. Bárány became interested in this phenomenon because it resembled labyrinthine nystagmus provoked by labyrinthine stimulation in that it consisted of a rapid phase in the direction of the moving train and a slow phase in the opposite direction. Studies on the physiology of nystagmus have been centered on the labyrinth, as the history of Bárány shows. Since the time when nystagmus was discovered to develop as a result of labyrinthine stimulation, it was considered to be an important objective expression of labyrinthine function. Bárány paid attention to railroad nystagmus because it resembled labyrinthine nystagmus. However, what is important is that nystagmus originally develops optically, as proven by phylogenesis. Previously, I reported on nystagmus using a *Crustacean*, the crab, as a model (Fukuda, 1957). The equilibrium organ of the crab is composed of only the statocyst, without a semicircular organ. When vision is blocked, the crab develops neither per-rotation nor postrotation nystagmus, although it shows optic nystagmus. This suggests that optic nystagmus should be treated as a primary phenomenon rather than as auxilliary to labyrinthine nystagmus, as in the traditional approach to this problem. In other

160

words, the labyrinth came to participate in nystagmus only after the statocyst evolved into the semicircular organ during evolution. (See Figures 14.9 and 14.10 in Chapter 14.)

As a natural result of physiological studies on equilibrium starting with the labyrinth, we have come to understand the significance of vision and the important role that equilibrium function plays in relation to receptors of the entire body. I hereby discuss the relationship between vision, or optic nystagmus, and posture. The German word *Haltung* for posture means not only the so-called posture constructed by the skeletal muscles, but also the position of the eyeballs inside the orbits.

There are a few reports in the literature on vertigo as well as motor ataxia provoked by optic stimulation. Fischer and Babcock (1919) mentioned that optic nystagmus as well as per-rotatory vertigo were produced by looking at objects moving successively in front of the eyes. He also described what he called *optokinetischer Körperreflex* (or optokinetic body reflex), in which the head and trunk were deviated toward the direction of the slow phase of nystagmus. Mach (1875) noted that when one looks down from a bridge at a river flowing beneath, the flow of the water suddenly appears to stop and at the same time one feels as if one were moving upstream with the bridge. These are simply descriptions of phenomena. I would like to move beyond them and evaluate the way animals, which stand against gravity, maintain body equilibrium while adapting to various stimuli such as light, sound, and pressure, and I will emphasize the role of vision.

I will show that just as excessive stimulation of the labyrinth results in experimental nystagmus as well as motor ataxia and vertigo, extraordinary optic stimulation produces optic disturbance in equilibrium. I will also describe how the changes in position, or *Haltung*, of the ocular and skeletal muscles affect each other.

A large rotating cylinder: A large rotating cylinder of paper 2.0 m in diameter and 1.7 m in height can be used for observing optic nystagmus (Fig. 12.1). It is hung from the ceiling of a well-lit room by a rope 2 m in length with the bottom of the cylinder 45 cm above the floor. Inside the paper cylinder there are 16 evenly spaced red vertical lines of 5 cm in width. The cylinder is rotated manually so that the rope is wound up and is then allowed to unwind at a constant speed of one rotation in 10 seconds. Revolution of the cylinder clock-wise (as seen in Fig. 12.1) is defined as rightward rotation, and the reverse leftward rotation. A circle of 2.0 m in diameter coinciding with the circumference of the cylinder is drawn on the floor under the cylinder. In addition, a concentric circle of 1.0 m in diameter is drawn on the floor, and the circles are divided by radial lines at 30-degree intervals.

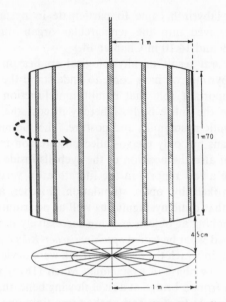

FIG. 12.1.   Large rotating cylinder. The illustration shows lines on
the outer surface of the cylinder, but in fact, 16 red
lines 5 cm in width are drawn on its inner surface (see
page 259 for a photograph of an actual cylinder). Optokinetic stimulation is given to subjects (or test animals
placed on a stand) in the center of the cylinder so as
to provoke optic nystagmus.

The subject stands or assumes various postures at the center of this
rotating cylinder (Fig. 12.2). The subject sees only the inside of the rotating cylinder, i.e., the moving red lines. Marked optic nystagmus is
seen to occur. Using the standing position as the fundamental position,
various experiments were made by employing the postures taken during
the stepping test, the stepping test with an otogoniometer, and the half-
rising posture.

Body equilibrium of normal subject: Before we discuss the main subject, the body equilibrium of a normal person standing inside the rotating cylinder should be mentioned. In summary, the rotation always provokes optic nystagmus at an average frequency of 64 times in 20 seconds
for both rightward and leftward rotations. When the subject marks time
with his eyes open while receiving this optic stimulation on the retina,
he is unable to retain the same straight position: his head, trunk, and
arms and legs deviate in the direction of rotation of the cylinder. Simultaneously, deviation and ataxia develop in such a way that the subject

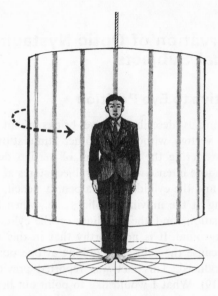

FIG. 12.2.   Standing position.

moves from the center towards the revolving cylinder. A subject even capable of assuming the straight standing position on one leg with eyes closed for more than 30 seconds will develop optic nystagmus as soon as he receives optic stimulation from the cylinder on the retina. Meanwhile Romberg's phenomenon takes place in the direction of rotation; the subject becomes unable to keep standing on one leg and puts the lifted foot down on the floor.

Strong vertigo is also provoked at the same time as this motor ataxia. The vertigo is such that the subject feels as if his body were rotating in the opposite direction to the rotation of the cylinder and as if he were attracted to the surface of the cylinder. The subject may also feel as if the paper surface were coming towards him—an illusion. In other words, concomitant with the optic nystagmus, strong motor ataxia and vertigo are produced. This phenomenon resembles the situation in which the labyrinth, which maintains body equilibrium, causes strong motor ataxia and vertigo when stimulated by 10 turns in 20 seconds, or when caloric nystagmus is provoked. Thus the fact that the labyrinth, which is an important receptor for equilibrium, can disrupt equilibrium when it receives unusual, excessive stimulation (what we call the "stage of disturbance" caused by excessive stimulation of equilibrium organs) also holds true in the case of optic stimulation.

12.2. Observation of Optic Nystagmus in Normal Subjects

12.2.1. Relative to Eye Position

The rotation stimulus described above brings about marked, purely optic, rotational vertigo without any other stimulation; this is an interesting fact considering the mechanism of vertigo development. This vertigo becomes more intense when the subject stares at an object placed between his eyes and the cylinder (e.g., a pen, a pencil, or a finger tip), instead of looking at the moving red lines; i.e., when his eyes are fixed in a straight position, and the red lines on the cylinder move successively in the background. It is noteworthy that in this case the eyeballs do not move, because the subject is looking at one point, and no optic nystagmus is produced. This fact was reported previously by Fischer and Babcock (1919). What I would like to point out here is that there is deviation in stepping when staring at one point. The angle of deviation in marking time is almost twice as great when the subject stares at the fixed point as when he looks at the rotating cylinder.

To substantiate this finding, we performed an experiment combining the cylinder with use of an otogoniometer. As shown in Figure 12.3, the eyes were kept straight relative to the head by making the subject wear

FIG. 12.3. Otogoniometer. Using this device, an examinee's eyes can be fixed in the median position.

FIG. 12.4.   Stepping test with otogoniometer in large rotating
cylinder.

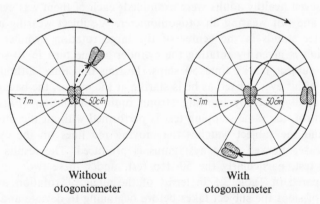

Without   With
otogoniometer   otogoniometer

FIG. 12.5.   Schematic presentation of results of stepping test in large
rotating cylinder with and without otogoniometer. The
left side shows the locus of a subject without otogoniometer, in whom optic nystagmus developed; the right
side shows that of a subject with otogoniometer, and
without optic nystagmus.

TABLE 3.5. Optic Equilibrium Ataxia during Stepping Test
with and without Otogoniometer

	Optic nystagmus	No. of steps taken before bodily rotation*	Angle of deviation
Without otogoniometer	+	24 steps	62°
With otogoniometer	−	14 steps	126°

*Number of steps taken before the subject started to rotate around the long axis
of his body in the direction of cylinder rotation.

TABLE 3.6. Subjective Ratings of Optic Vertigo with Otogoniometer

Effect of otogoniometer	Male $N=35$	Female $N=16$	Total $N=51$
Vertigo increased	30 (85.7%)	13 (81.3%)	43 (84.3%)
Vertigo unchanged	2 (5.7%)	2 (12.5%)	4 (7.8%)
Vertigo decreased	3 (8.6%)	1 (6.3%)	4 (7.8%)

an otogoniometer, a device we frequently use when observing experimental labyrinthine nystagmus. The subject inside the rotating cylinder
stares at his eyes reflected in a small mirror placed 30 cm in front of
him, so that his eyes are constantly positioned straight in relation to his
head while performing the stepping test or other tests.

Sixty-seven healthy adults were examined; each of them was evaluated
wearing and not wearing an otogoniometer. A subject wearing an otogoniometer stands in the center of the large rotation cylinder. Optic
stimulation is given by rotating the cylinder at one turn in 10 seconds,
and the subject marks time 50 steps. During this time, although the
position of his eyes is fixed as he is staring at the mirror, the background
moves rapidly, thereby exerting strong optic stimulation (Fig. 12.4).
As a control, the same stepping test is performed without the otogoniometer, and the subject watches the moving red lines on the cylinder.
In this case, marked optic nystagmus is produced. The results of the
stepping test, particularly the 50-step test, under these two conditions
are compared in Table 3.5 in terms of the angle of rotation and the
number of steps the subject takes before beginning to deviate and rotate
in the direction of the rotation of the cylinder. Figure 12.5 is a schematic
presentation of the results. The results show that although the subject
received the same stimulation, his ataxia was nearly doubled by wearing
an otogoniometer. It is also noteworthy that with the otogoniometer
the motor ataxia of deviation and rotation in the direction of rotation
of the cylinder were developed after half as many steps as without the
otogoniometer. Table 3.6 shows the subjective ratings of optic vertigo

of examinees with and without the otogoniometer; optic vertigo was produced by the excessive stimulation of rotation of the large cylinder. Eighty-five percent of the examinees claimed that vertigo was markedly increased by wearing the otogoniometer.

In summary, when optic nystagmus was suppressed by staring at an otogoniometer while the cylinder was rotating, the resulting equilibrium ataxia was about twice as great as when active optic nystagmus was not suppressed. Moreover, optic vertigo was increased in most examinees (about 85%) by the otogoniometer. Although in the previous studies of optic vertigo, emphasis was placed on subjective sensations, what we showed clarified that this vertigo was always associated with ataxia of the skeletal muscles. In our experiment, we showed that the more optic nystagmus was suppressed and the more marked ataxia of skeletal muscles became, the greater was the optic vertigo. This parallel relation indicates that the objective manifestation of vertigo is ataxia of skeletal muscles. In this way, we clarified the relationship among optic nystagmus, optic equilibrium ataxia, and optic vertigo.

12.2.2. Relative to Body Position

It is well known that nystagmus and change in the tension of skeletal muscles develop when marked stimulation is given to the labyrinth or eyes. In the previous section, I described the change in the skeletal muscles when the eyes were fixed in the median position. In this section, I will analyze the change in the eye muscles or nystagmus caused by change in skeletal muscles produced by optic stimulation. I will also show that there is a relationship between posture and optic nystagmus and that the subjective visual field is greatly influenced by posture.

The subjects studied were ten healthy untrained people, six professional baseball players, and five sumo wrestlers. Both baseball players and sumo wrestlers were top-ranked athletes. They were cooperative in the objective study of good athletic abilities (or dynamic equilibrium capacities), which constitutes part of equilibrium physiology.

Inside the rotating cylinder the subjects adopted postures that are basic in baseball or in sumo wrestling:

1) Standing position: The subject stands in the center of the cylinder with the tips of his toes together facing forward and his upper limbs hanging naturally (Fig. 12.2). He first closes his eyes and then opens them so that the constantly moving red lines on the inner side of the rotating cylinder are projected onto his visual field, causing nystagmus. This nystagmus is observed for 20 seconds and its frequency is counted.

2) Half-rising posture: As shown in Figure 12.6, the subject stands

with his feet apart and bends down, placing the palms of his hands on his knees. In this position the upper part of his body is bent slightly foward. Sumo wrestlers assume this posture before they begin to take *shiko* steps.

3) Sumo-squatting posture: In this position, the trunk is lowered more than in the half-rising position and the subject places his clenched fists on the ground. This is the position from which sumo wrestlers spring up to fight—a position filled with tension (Fig. 12.7).

4) Baseball batting stance: The baseball players assumed their ordinary standing position (Fig. 12.8) as well as their favorite batting stance, although not with a bat.

The results of examinations in these four postures are tabulated in Tables 3.7, 3.8, and 3.9. They show that optic nystagmus was greatly influenced by the posture of the examinees.

Table 3.7 shows that in eight of the ten non-athletes there was less optic nystagmus in the half-rising position than in the standing position, the maximum difference being 25 during 20 seconds of optic stimulation. The frequency of nystagmus was the same in the two positions in one subject (N10) and greater in the half-rising position in another. Similar results were obtained with sumo wrestlers in their postures of standing, half-rising, and sumo-squatting (Table 3.9). Subject S1 had a frequency of 57 while standing and 41 while sumo-squatting. Baseball players also showed less nystagmus when their bodies were bent (Table 3.8). Subject B3 had a frequency of 73 while standing and 55 while assuming the batting stance.

All the participants in this examination noticed an interesting phenomenon of a subjective nature. They all claimed that the more clear and distinct the moving red lines on the cylinder appeared, the wider was the range of their visual field and the less was their vertigo when assuming the half-rising, sumo-squatting, or batting stance. Members of my research group also tried this test in various postures; we found that when the body was bent, the frequency of nystagmus was decreased; movement of the red lines was clearer; the range of the visual field was wider, particularly in the left to right direction; and the position could be maintained better because vertigo was less. These subjective sensations cannot be directly substantiated by objective reduction in frequency of nystagmus. The widening of the visual field has been substantiated by detecting widening of the nystagmus, which we did measure quantitatively, but we do not know how these lead to reduction in frequency of nystagmus.

As I have shown above, there is a relation between optic nystagmus and posture. Now I would like to discuss the question of the small of the back, which is considered very important in any kind of sport. When baseball players in the field are waiting for a ball to come, they never

FIG. 12.6   FIG. 12.8

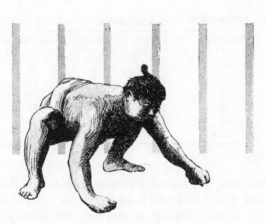

FIG. 12.7

TABLE 3.7. Frequency of Optic Nystagmus in Ten Non-Athletes (20 sec.)

Subject	Standing position (A)	Half-rising position (B)	Difference (A−B)
N1	56	37	19
N2	62	42	20
N3	66	53	13
N4	60	35	25
N5	62	64	−2
N6	67	53	14
N7	62	59	3
N8	62	48	14
N9	55	49	6
N10	57	57	0
Average	60.9	49.7	11.2

TABLE 3.8. Frequency of Optic Nystagmus in Six Baseball Players (20 sec.)

Subject	Standing position (A)	Half-rising position (B)	Difference (A−B)	Batting stance (C)	Difference (A−C)
B1	64	56	8	41	23
B2	60	54	6	62	−2
B3	73	71	2	55	18
B4	52	41	11	36	16
B5	70	52	18	40	30
B6	42	36	6	40	2
Average	60.2	51.7	8.5	45.3	14.9

TABLE 3.9. Frequency of Optic Nystagmus in Five Sumo Wrestlers (20 sec.)

Subject	Standing position (A)	Half-rising position (B)	Difference (A−B)	Sumo-squatting position (C)	Difference (A−C)
S1	57	56	1	41	16
S2	71	62	9	65	6
S3	65	59	6	56	9
S4	63	58	5	63	0
S5	66	59	7	56	10
Average	64.4	58.8	5.7	56.2	8.2

stand like spectators, but assume a posture such as that shown in Figure 12.8; they bend their backs and keep their eyes straight forward. This is also true in tennis. Whether in receiving fast services or volleying quickly, players usually assume the half-rising stance. In this posture, the width of nystagmus is increased and its frequency is reduced, as we established objectively. In this way players see moving objects definitely and clearly with a widened visual field. The sumo-squatting position allows the wrestler to see the opponent's slight movements clearly in the widest possible visual field, and to be able to respond to the opponent's strategy. Thus, I claim that, on the basis of the above considerations, the small of the back is significant in sports in connection with the relationship between optic nystagmus and posture.

12.3. Discussion

From the standpoint of optic nystagmus, I have shown that changes in eye position and the posture constructed by the skeletal muscles lead not only to development of optic ataxia and vertigo, but also to changes in the area of vision. In view of this, I would like to consider equilibrium function and ataxia. Vision is as important as the labyrinth for equilibrium physiology. Ataxia in a deaf person with impaired labyrinthine function is compensated for by vision, so that his equilibrium is almost completely normalized. This is illustrated in Figure 12.9. The photos on the left show the posture of a boy without labyrinthine function. When his eyes are open, he can maintain balance on an inclination. He moves his trunk to counteract the effect of the inclination; that is, he assumes a normal position in relation to gravity thereby preserving equilibrium. When his eyes are covered, his ability to maintain equilibrium is entirely lost, as shown in the pictures on the right of Figure 12.9, and he falls in the direction of the inclination. In short, he maintains equilibrium solely by vision. It is important to note that this is achieved through vision without any conscious effort to balance. This indicates how important vision is in maintaining equilibrium. When unusual optic stimulations not encountered in daily life are applied successively to the retina, as in our rotating cylinder, not only optic nystagmus but also optic motor ataxia of the skeletal muscles is provoked together with marked vertigo. Paleness, cold sweat, nausea, and other symptoms of ataxia of smooth muscles may also be produced. In summary, as I mentioned in relation to the labyrinth, there are two phases of response in optic equilibration—a phase of response that maintains equilibrium on appropriate stimulation and a phase of disruption of equilibrium on excessive stimulation.

   Phenomena in the body are quite mysterious in that the labyrinth and vision are both important organs for maintenance of equilibrium, but both disrupt balance when stimulated excessively. The same is also true for other organs of the body. For instance, two opposite reactions are seen in the stomach: normally food accumulates in the stomach for further digestion, but when an abnormal quantity or quality of food is eaten,

FIG. 12.9.   A deaf boy with no labyrinthine function on a slope.
Left: with his eyes open, he can maintain body equilibrium against inclination of the board. Right: with his
eyes closed, he loses balance at even a slight inclination
of the board.

it is vomited from the stomach. What I wish to emphasize from these
facts is that biological function can manifest opposite responses depending on the quality and quantity of stimulation, so the entire biological
function of an organ cannot be deduced from one fixed phenomenon.
Phenomena may be opposite depending on the quantity of stimulation,
but they should be understood as coming from the same stream.

Last, I wish to consider the reflex in relation to equilibrium. A phenomenon which appears in one part of the body in response to stimulation of another part is loosely termed a reflex. However, as I have repeatedly mentioned, labyrinthine or optic stimulation results in two phases—a phase of response or coordination, and a phase of disturbance or disruption of equilibrium. The phenomenon described above of maintaining balance on a slope by means of vision when labyrinthine function is impaired is called an optic equilibrium reflex. Motor ataxia and vertigo during experimental labyrinthine nystagmus or optic nystagmus are also called reflexes. Romberg's phenomenon is a representative example.

I do not mean to say that the term reflex should be limited to reactions that are expedient. However, I feel hesitant in making no distinction between the phase of response, during which equilibrium is maintained, and the phase of disruption, in which equilibrium is lost because of excessive stimulation. Optic nystagmus can be considered to be a reflex effecting equilibrium maintenance: when it was suppressed in the experiment using an otogoniometer, motor ataxia of the skeletal muscles nearly doubled. I feel that it is quite inconsistent to call the ataxia of skeletal muscles brought about by labyrinthine stimulations a labyrinthine spinal reflex; *Diskuswerferstellung*, which is a marked kind of Romberg's phenomenon, is one example of this. I also feel that calling the optic motor ataxia produced by the rotating cylinder simply an optic spinal reflex is inconsistent from the standpoint of equilibrium function. I do not think that it is appropriate to use the same term for a phenomenon that is expedient for maintenance of life and also for a response of one part of the body produced via nerves by stimulation of another part of the body.

I would like to mention here that the phase of disruption can be suppressed and that the phase of response can be extended by training. Birds and animals acquire through daily training the ability to withstand excessive stimulation without developing ataxia. They can develop this ability through excessive stimulation. The cerebrum and cerebellum play important roles in this development: it is not just a conditioned reflex. This is something of special interest to us.

Part 4

The Position of the Labyrinth in Equilibrium Function

Chapter 13

Memorandum on the Vestibular Labyrinth

13.1. Vulnerability of the Vestibular Organs

Since the work of Wittmaack (1911), vulnerability of the vestibular and cochlear nerves has been frequently discussed, and it has been concluded that the former is less vulnerable than the latter. Although there are various experimental findings bearing on this idea, it first arose from the clinical opinion that "the equilibrium organ is less vulnerable than the hearing organ"—an opinion based on unclear and unresolved premises.

It is certainly true that various ear diseases are associated with hearing impairment—even in otitis externa hearing impairment can develop—and, conversely, equilibrium impairment is rarely accompanied by disease. In other words, although mild ear diseases can be associated with impairment of hearing, vertigo or equilibrium impairment seldom develops. It seems to me that the simplistic conclusion that the vestibular nerve is less vulnerable than the cochlear nerve was established on the basis of these clinical observations.

Hearing impairment can be classified into two types: conductive and perceptive impairment. This classification was made by our predecessors in otolaryngology from clinical experience and autopsy findings. Based on this classification, hearing impairment is analyzed in a reverse fashion to locate the exact site of a lesion. That is, diseases of the external ear and middle ear give rise to conductive hearing impairment, whereas those of the cochlea result in perceptive hearing impairment. Thus it can be deduced that if conductive impairment is present, the lesion does not extend beyond the middle ear and that the cochlea and more central area are unaffected.

The conductive organs (external and middle ear) are amplifiers of sound that are unrelated to nerves. Examination of the cochlear nerve in conductive hearing impairment leads to the conclusion that the cochlear nerve is intact and unimpaired.

Therefore, when discussing the vulnerability of the vestibular and cochlear nerves clinically, equilibrium impairment should be compared with perceptive hearing impairment, not conductive impairment.

When thinking in this way, one finds that the assumption that the vestibular nerve is not easily damaged is not well substantiated. When otitis media spreads to become otitis interna, vestibular impairments, such as nystagmus, vertigo, and equilibrium disturbance, develop, and at the same time conductive impairment turns into perceptive impairment, as shown by a reversed result in the Weber test and a shortened bone conductive time. In short, the vestibular and the cochlear nerves are affected simultaneously, without any difference in vulnerability. In Ménière's disease, which is a pure inner ear disease whose pathology has not yet been fully worked out, vertigo, hearing impairment, and tinnitus develop, i.e., both vestibular and cochlear symptoms appear at the same time; the vestibular symptoms are more marked than the cochlear symptoms. When analyzing the vulnerability of the two nerves, which should be done most carefully with regard to inner ear diseases, it becomes apparent that there is no difference in vulnerability of the two, as shown by the above example. Thus I maintain that there is no ground for saying that the vestibular nerve is less vulnerable than the cochlear nerve.

Audiographic examinations often give results that are not clearly due to either conductive or perceptive disturbance, but to a mixture of the two. In such instances, not vestibular disturbance, but both conductive and perceptive disturbances are present. Middle ear diseases can often be diagnosed on the basis of changes in the tympanic membrane. This lesion of the middle ear affects the cochlea of the inner ear to cause the above kind of mixed hearing impairment. Association of tinnitus with a middle ear disease is clear proof that the cochlea is affected. It should be recognized clinically that a lesion that is limited to the conductive organs can exert some influence on the cochlear nerve. On the other hand, it is not usually recognized that vestibular function in not affected in a middle ear disease. Anatomically, however, just as the middle ear is located close to the cochlea, the vestibular organ lies next to the middle ear by way of the oval and round windows. The question arises of whether stimulation of the middle ear cavity gives rise to some unknown equilibrium impairment that is different from nystagmus or vertigo, just as a middle ear lesion causes hearing impairment. This is our first question to be answered.

13.2. Latent Nystagmus

In examining vestibular function, the labyrinth is usually strongly stimulated to produce disruption of body equilibrium, of which nystagmus is a representative marker. When this type of disturbance is provoked, it is concluded that a labyrinthine reflex has been developed; in the absence of this disturbance, it is concluded that no labyrinthine reflex is produced. (According to this interpretation, the labyrinth is defined as an organ causing dysequilibrium on stimulation, or an organ producing vertigo on stimulation. A labyrinth that does not give rise to vertigo is diagnosed as afunctional or dysfunctional. This definition leads to the conclusion that the equilibrium organ is a vertigo-producing organ. Thus this is a very strange definition; I will discuss this in more detail later.) When the position of the head is altered, the position of the eyeballs changes; even on slight rotation, rotatory nystagmus develops. In this way, on the slightest static or dynamic labyrinthine stimulation, the ocular muscles respond sensitively and restore equilibrium. The reflex of the skeletal muscles has not been fully analyzed. It would be interesting to know objectively how the skeletal muscles respond to a degree of labyrinthine stimulation insufficient to cause artificial nystagmus.

There is a concept of latent nystagmus, or *Nystagmusbereitschaft*. This vaguely means a condition in which there is something unusual about labyrinthine equilibrium, but not so much as to cause clearly objective symptoms such as vertigo or nystagmus. Our second question is what objective findings are presented in this poorly defined condition of latent nystagmus.

13.3. Two New Methods for Testing the Vestibulospinal Reflex

Romberg's method is most often used to test the vestibulospinal reflex. In this method, the subject stands with his eyes open or closed, and the direction in which his body sways or falls is observed. When nystagmus is present during this test, the body sways and falls in the direction of the slow phase of nystagmus, as one would expect. When labyrinthine disturbance is suspected even though the results of Romberg's test are negative, another test is given with the subject adopting an unstable posture or performing certain movements; i.e., the subject stands on one leg or on a slope, or walks with his eyes open or closed to test for equilibrium disturbance that would not be observable by Romberg's method. The labyrinth forms a reflex arc in the medulla oblongata where close connection is made with muscles of the body. In trying to make a detailed analysis of the spinal reflex in muscles, Bárány devised the past-pointing test of the upper limbs. The walking test is used to examine the lower limbs. In general, the labyrinthine reflex appears more clearly when the muscles are smaller and more movable. Ocular muscles, which are small and very movable, produce the reflex sensitively. In man, small muscles other than the ocular muscles do not produce an overt reflex, but in small animals, rolling of the head or the body is observable. In order to make the spinal reflex overt in man, vision is blocked, so that the optic reflex for compensation of labyrinthine disorders is eliminated, and the subject then performs movements in postures as unstable as possible in relation to gravity to make labyrinthine deviation manifest. This deviation is imbalance in the labyrinthine reflex that participates in the volitional control of muscles by the pyramidal tract. In this way, latent imbalance in labyrinthine muscular tension becomes overt. A characteristic of labyrinthine equilibrium disturbance is that it has directionality. Except in cases of bilateral absence of labyrinthine function, deviation in a certain direction produces positive results in Romberg's test, standing on one leg, standing on a slope, or walking, because maintenance of balance is poor in a certain direction. In this way, deviation is important in examining labyrinthine equilibrium. Man lives in three-dimensional space, and his skeletal muscles move freely in these three dimensions. Therefore, deviation in equilibrium disturbance is best analyzed in terms of the anteroposterior, superoinferior, and dextro-sinister planes and the superoinferior, frontal, and sagittal axes. Bárány's pointing test is based on this concept.

The blindfold writing test: Previously I proposed the blindfold writing test as a new test of the spinal reflex. The main idea can be summarized as follows. The subject writes a series of letters or symbols from top to bottom on a paper with his eyes open and then with them closed. Motor ataxia that interferes with writing movement becomes obvious when vision is blocked and can be detected by comparing the writing with eyes open and closed. In order to observe pure motor ataxia as nearly as possible, the subject is made to write with the tip of the pencil, keeping his arm above the desk. Bárány observed deviation in movement of the shoulder joint, as well as in the elbow, wrist, and finger joints; he indicated that the muscles that move these joints all show the same difference in muscular tension. Examination of the small joints below the shoulder is usually difficult, but all the muscles of the upper limb are involved in the movement of writing with eyes closed.

The delicate, anteroposterior and left-to-right movements of writing involve the small muscles of the arm and finger joints. The deviation and other forms of motor ataxia indicated by the series of written letters are valuable and significant in the sense that changes in the tension of the upper limb muscles are expressed all together. Usually during this test, the paper is placed on a desk, and this allows evaluation of the deviation in the left to right and anteroposterior directions. If the paper is placed vertically, deviation in the superoinferior direction can also be evaluated.

The stepping test: The walking test has frequently been performed. In this test, the subject walks a certain distance on a straight line with eyes closed, while deviation and performance in walking are evaluated. Normally, the subject walks six meters, and deviation of one meter or more laterally is considered to be significant. However, it has been frequently deplored by researchers that the outcome of this test is not consistent, and so this method is used only as an auxiliary test at present. The shortcomings of this method are as follows. First, the subject walks only a few steps, which is not sufficient to allow deviation to become clearly manifest. Second, the subject becomes worried about moving laterally and bumping into things, which hinders his natural walking; in fact, he often stops walking in the middle because of this worry. Third, attention is paid only to lateral deviation: anteroposterior movement, and rotation around the long axis of the body are not evaluated in this test. Accordingly I devised the stepping test, which does not have these shortcomings. This test is briefly as follows. The examinee, with eyes closed, stretches both arms straight out in front of him and marks time 50 or 100 times at the center of concentric circles of 0.5, 1.0, and 1.5 m in diameter which are divided into 30-degree radial segments. The deviation produced while marking time is evaluated in relation to the function of the legs. In this test, the subject can mark time an unlimited number of times in a small space, and he does not worry since he thinks he is remaining in the same place, whether he has actually deviated or not. His body moves with deviation of the lower limbs. In the pointing test and the writing test with eyes covered, deviation is limited to the range of joint motion, but in the stepping test, this limitation does not apply. Deviation is analyzed not only in the anteroposterior and left-to-right directions in this test, but also in the direction of the long axis of the body. In other words, as the subject marks time, his body can turn to the right or left around the long axis of the body. This movement should not be treated lightly, but regarded as an essential labyrinthine function. When the labyrinth of the rabbit is extirpated, the body rolls around its long axis towards the operated side. In humans, no such rolling movement is observed even in severe labyrinthine impairment. But in the stepping test, revolution of the body around the long axis is almost always observable, a most noteworthy finding. I interpret this revolving as the same phenomenon as the body rolling seen in animals, and I think that it is caused by a similar change in labyrinthine muscular tension as is the body rolling.

By applying various known labyrinthine stimuli, this phenomenon can be provoked clearly. Therefore, it is attributable to the change in muscular tension induced by the labyrinth. In this regard, the stepping test should not be thought of as a modification of the walking test, in which only lateral movement of the body is observed, because in the stepping test not only movements in the anteroposterior and left-to-right directions but also body rotation around the long axis are examined.

New facts are discovered by new methods of examination. The two questions posed in the previous sections have been solved by the writing test with eyes covered and the stepping test, thereby clarifiying an unknown area of labyrinthine functions.

13.4. Effects of Middle Ear Diseases on the Labyrinth and Provocation of Latent Nystagmus

When I examined normal people and patients with ear diseases using the blindfold writing test which I had devised, I obtained some strange results. Unexpectedly I found that patients with various middle ear diseases showed marked deviation in the writing test, but that this was never seen in patients with normal otologic function. Patients with otitis media showed marked deviation in writing, which disappeared after treatment. Patients with a unilateral stenotic Eustachian tube also often showed marked writing deviation, and when the passage was reestablished, this deviation disappeared immediately and they could write in a straight line. The examinees stated that this was very mysterious. Although reestablishment of the tubal passage was conducted with some expectation of improvement in results in the writing test, the finding was nonetheless surprising. The fact that the deviated writing was corrected by restoration of the tubal passage indicates that the origin of the phenomenon lies in the ear. In other words, the stenosis of the middle ear cavity on one side affects the labyrinth and creates latent imbalance in labyrinthine muscular tension, which causes deviation in writing. In these patients, no laterality in nystagmus or deviation in the pointing test was provoked experimentally. In other words, traditional examinations failed to detect such a delicate difference in the labyrinthine muscular tension (cf. Fig. 5.7 in Chapter 5).

In various examinations, one often finds it difficult to make a clear distinction between what is normal and what is pathologic, except when the result of an examination turns out clearly black or white. Therefore, statistics become important; the normal range is determined and values outside the normal range are considered to be abnormal. However, many values fall in an intermediate zone, which could be normal or abnormal. In the cases of unilateral tubal stenosis in which deviated writing became quite normal when the passage was reestablished, this trouble did not arise. The deviation was undeniably caused otologically by labyrinthine involvement. This fact gives the answer to the question of whether stimulation of the middle ear cavity gives rise to some unknown equilibrium impairment that is different from nystagmus or vertigo. In short, just as middle ear diseases affect the cochlea, they also affect the labyrinth and bring about latent imbalance in the labyrinthine muscular tension. What sort of results are expected in the stepping test, then? Two examples are given in Table 4.1.

Otologically normal persons do not show deviation in the stepping test. The maximum deviation that is considered to be within normal limits is 30 degrees to either side (or 60 degrees from right to left) at the 50th step, or 45 degrees to either side at the 100th step. But otologically normal persons can develop marked deviation resulting in rotation around the long axis of the body when they receive labyrinthine stimulation. An interesting observation is that when weak labyrinthine stimulation—such as 2 turns in 10 seconds, infusion of 1 ml of water at 27°C into the ear, or increase or decrease of the external auditory meatus pressure, which do not cause an appreciable nystagmus—is given, marked deviation and rotation around the long axis far exceeding the normal range are clearly demonstrated in the stepping test. In other words, the labyrinthine spinal reflex is clearly demonstrated even when no nystagmus is provoked experimentally. This answers our second question of whether latent nystagmus or a state of imbalance in labyrinthine muscular tension exists: this state was demonstrated objectively.

When changes such as inflammation, secretion, or mucous thickening are present in the middle ear cavity, the tympanic membrane is swollen and the stapes at the oval window and membrane of the round window are affected. This may exert a pressure on the inner ear. If the malleus happens to be pushed inward in Eustachian stenosis, the stapes which is connected to the malleus is then pushed inward, raising the inner ear pressure. If for some reason a pressure difference is produced between

TABLE 4.1.   Results of Stepping Test in Patients with Middle Ear Diseases

---

1.   Right acute purulent otitis media in a 40-year-old male

Case history: Right-sided otalgia and discharge had started four days previously for no obvious reason. The discharge had stopped the previous day, although impaired hearing and tinnitus persisted.

Physical status at the time of consultation: Redness and swelling of the postero-superior wall of the body part of the external auditory meatus was observed. The tympanic membrane was diffusely red and thickened with marked redness in the posterosuperior quadrant. No headache or vertigo was present.

Stepping test: Posture while marking time was steady, although the body turned toward the affected side, i.e., to the right, at each step. Simultaneously, the body gradually moved forward to the right. Body rotation became more and more marked during the test: by the 50th step, the body had turned 120 degrees, and by the 100th step, 360 degrees. In other words, although the patient thought he was marking time in exactly the same place, he had actually made one complete revolution to the right.

---

2.   Right chronic otitis media catarrhalis in a 30-year-old male

Case History: Five months previously, the patient developed otalgia and otorrhea of the right side, which were alleviated by medical treatment. Then he began to complain of a sense of right ear obstruction and difficulty in hearing.

Physical status at the time of consultation: A line of exudate and bulla formation was observed on the right tympanic membrane. The hearing impairment in the right ear was a typical conductive disturbance.

Stepping test: No swaying of the body was observed during the test, and body movement was steady. However, when the patient marked time, his body gradually turned to the right. By the 50th step, he had turned 180 degrees; i.e., he was facing in the opposite direction. At the 100th step, he had turned 450 degrees, i.e., 1 1/4 rotations. After air douche, the exudate line and bulla disappeared and hearing was restored. On repeating the stepping test after this treatment, he moved slightly forward, but did not revolve; i.e., he was able to remain facing forward while marking time 100 times.

---

the external auditory meatus and the middle ear so that the pressure of the middle ear is higher, thereby making the tympanic membrane swollen, the pressure of the inner ear will be changed. The labyrinthine stimulation gives rise to deviation in stepping in patients with middle ear disease is probably produced by the change in pressure in the inner ear which is created by the change in that of the middle ear. Formerly, no labyrinthine reflex caused by change in pressure in the external auditory meatus was known, although the fistula symptom, i.e., the nystagmus produced by increase or decrease in the external auditory meatus pressure, was known. The spinal reflex caused by the pressure change in the external auditory meatus of the normal ear was not known previously. I will describe this reflex or deviation in stepping below on the basis of the experimental findings of Yoshikawa and Kato.

If positive or negative pressure is applied to the external auditory meatus while performing the stepping test, definite rotation around the long axis and lateral movement are produced in persons with normal ears. When positive pressure of $+70$ to $+100$ mmHg is applied, deviation is produced in the direction of the pressure, whereas when negative pressure of -50 mmHg is applied, deviation is in the opposite direction. When a rubber tube with a blind end is inserted into the bony part of the external auditory meatus in such a position that it presses on the tympanic membrane without causing pain, this pressure on the tympanic membrane results in marked deviation in stepping in the direction of the ear with the rubber tube.

In other words, when pressure is applied with the rubber tube in the external auditory meatus or directly to the tympanic membrane of a normal person, it creates the same condition in the middle ear and inner ear as that in otitis media or tympanic membrane depression. This observation clarifies the mechanism of deviation in stepping of patients with middle ear disease. The fact that the direction of deviation was the same in this experiment as in the patients supports this conclusion.

We have thus confirmed clinically and physiologically that middle ear disease can affect the labyrinth and bring about latent imbalance in muscular tension via the labyrinthine spinal reflex. These findings disprove the traditional concept that the vestibular nerve is less vulnerable than the cochlear nerve. Just as the cochlear nerve is affected in middle ear disease, the labyrinth is affected in middle ear disease, thereby producing latent imbalance in muscular tension, as proven by our new methods, the blindfold writing test and the stepping test. We have also clarified the concept of latent nystagmus which has hitherto been considered only vaguely.

13.5. What is Vestibular Function?

Although we are quite aware that the labyrinth is the organ of equilibrium, we examine its function mainly by examining nystagmus and vertigo, which we produce artificially. Some workers even go so far as to think that unless nystagmus is produced, the labyrinth has not been stimulated, or that the labyrinth is stimulated only after the development of nystagmus. This idea may be somewhat useful clinically, but must be discarded when studying labyrinthine physiology. The reason is as follows. If it is assumed that unless nystagmus, and therefore vertigo, is produced, labyrinthine function is impaired or destroyed, then the labyrinth must be regarded as a vertigo-producing organ. On rotatory or caloric stimulation, the labyrinth certainly does produce vertigo and disrupts body equilibrium, but if this argument is carried to its extreme, the labyrinth must be regarded as valueless because it produces vertigo which is unnecessary for man. In the previous section, I referred to various phases of the labyrinthine reflex produced by weak and strong stimulation. I believe that the true function of the labyrinth lies in the reflex produced by weak stimulation. Nystagmus and vertigo are provoked when the labyrinth receives unusually strong stimulation; the labyrinthine reflex produced by mild stimulation is quite different. In the blindfold writing test, completely opposite muscle reflexes are produced when there is marked postrotatory nystagmus and when there is not. When the subject is made to mark time after slow rotation of 2 turns in 10 seconds, which produces hardly any observable nystagmus, the body turns unconsciously in the direction opposite to rotation. In contrast, after 10 turns in 20 seconds, while nystagmus is still present, the subject staggers and gradually begins to turn in the direction of rotation. Thus the direction of turning is opposite to that after weak stimulation. However, when the nystagmus after strong stimulation ceases, the body starts to rotate in the opposite direction. In other words, the same results are obtained in the stepping test when the subject is given weak rotation stimulation and when his after-nystagmus from strong rotation stimulation disappears.

Previously, it was thought that deviation in the spinal reflex after rotation was in the direction of rotation. Rising after 10 turns in 20 seconds, the subject tends to fall in the direction of rotation, and the same deviation after rotation is seen in the pointing test and the stepping test. This was understood to be the labyrinthine spinal reflex, and no one doubted this conclusion. However, the deviation in the stepping test after slow rotation, which is opposite to the direction of rotation, is not consistent with the traditional concept of the spinal reflex. How can these two phases of deviation, which are produced by slow and fast rotations and are in opposite directions, be explained? Incidentally, the same results can also be obtained when the subject stands on one leg after rotation at low and high speeds. Further details will not be given here, but this shows that this kind of deviation with two phases is not restricted to the stepping test.

Generally speaking, any object that is rotating is subject to a centrifugal force. When rotation is stopped suddenly, the object deviates and falls in the direction of rotation; not in the opposite direction to rotation. After 10 turns in 20 seconds, Romberg's phenomenon, which is in the direction of rotation, and deviation in the pointing test are affected very much by centrifugal force and do not necessarily have to be ascribed to the labyrinth. In other words, they do not need a specific labyrinthine spinal reflex to account for them. To attribute Romberg's phenomenon or falling in the direction of rotation to a labyrinthine reflex is like attributing the fact that a passenger falls in the direction of movement when a train suddenly stops to a labyrinthine reflex. Teleologically speaking, when the body is subject to a centrifugal force on rotation, the labyrinth via the spinal reflex induces muscular tension that opposes the centrifugal force so that the body equilibrium is maintained against it. In the stepping test, the deviation after rotation that develops in the direction opposite to that of rotation can be thought of as a manifestation of this teleological labyrinthine reflex. In contrast, the deviation caused by the so-called spinal reflex after strong stimulation of 10 turns in 20 seconds is brought about by a change in muscular tension caused by the centrifugal force which is stronger than the effect of the labyrinth. The labyrinth thus gives in and the muscular tension changes. In other words, the deviation caused by strong stimulation should not be treated as a typical labyrinthine spinal reflex. I maintain that the true labyrinthine reflex is the one that produces muscular tension in the opposite direction to rotation.

Previously, caloric stimulation and pressure change of the external auditory meatus were given as stimulations, and artificial nystagmus was used as an indicator. It will be interesting to evaluate stimulations in the above-mentioned context. The vestibular nerve and fasciculus longitudinalis posterius are ontogenically and phylogenically the oldest. They are present in fish, birds, and other animals. It is interesting to compare results in the stepping test in man upon changing the pressure of the external auditory meatus to the behavior of a bird in the air. When wind pressure is applied to the side of the bird's body, the pressure on the external meatus on the windward side increases and that on the leeward side decreases. In the stepping test, deviation occurs toward the side of increased pressure and in the opposite direction to reduction in pressure. Thus, in the bird, muscular tension is probably increased via the labyrinth on the windward side, thereby allowing the bird to fly straight in spite of the wind pressure. On cold water infusion (27°C, 1 cc), deviation in the stepping test is toward the side of infusion, whereas on warm water infusion (48°C, 1 cc), it is toward the other side. This may explain why fish in water move toward cold water and away from warm water. Some instinctive phenomena like these may be ascribable to the labyrinthine reflex.

The above description may overstress the function of the labyrinthine reflex, but I believe that the essence of the labyrinthine reflex can be truly understood only by thinking along these lines. I do not approve of incessantly stimulating the labyrinth until nystagmus is produced. I cannot agree with the idea that the labyrinth has been stimulated and a reflex is provoked only when nystagmus is produced. In the traditional method of producing nystagmus, the stimulation was too strong, and thus the function of the labyrinth was disrupted. Observation was made in the phase of disruption. Before this phase, there is a phase of response or reflex in which equilibrium maintenance, i.e., essential labyrinthine function, is controlled, or skeletal muscle tension is adjusted. I feel strongly that the two phases of the labyrinthine reflex, those of disruption and response, should be recognized. The phase of response is not an abstract idea: we have demonstrated it objectively. These two phases appear differently depending on the individual and his physical condition. In very good athletes, dancers who are used to rotation, children who have been trained in rotation movement, and other persons who possess good labyrinthine function, postrotatory nystagmus appears sooner, and the frequency of nystagmus is less than that in normal persons after 10 turns in 20 seconds. According to the traditional idea, these people have reduced labyrinthine or equilibrium function, but this does not seem reasonable. In the stepping test, when good athletes mark time, the body turns and deviates in the opposite direction to rotation; that is, the phase of response of the reflex is seen. The traditional labyrinthine physiology is to me a nystagmology in which only the phase of dysequilibrium is emphasized. The main role of the labyrinth lies not in nystagmus or vertigo, or the phase of dysequilibrium, but in the phase of response, as I have reiterated.

Last, a few words on the labyrinthine autonomic nerve reflex. States of pleasure experienced while playing on a swing, see-saw, slide, or while dancing, skating, or skiing, all of which give labyrinthine stimulation, are in contrast to those of discomfort experienced in sea or air sickness. The former can be categorized as belonging to the phase of response of the labyrinthine autonomic nerve reflex, and the latter to that of disruption. As is well known, there have been conflicting findings on the labyrinthine autonomic nerve reflex. Some say respiration is suppressed and others say it is accelerated; gastrointestinal motility is either increased or reduced; blood pressure is either elevated or depressed. I suspected that these contradictions may be resolved by introducing the idea of two periods of response and disruption in studies on the labyrinthine autonomic nerve reflex.

Chapter 14

The Position of the Labyrinth in Equilibrium Function

14.1. The Term "Equilibrium Sense"

The term "equilibrium sense" has been used without much consideration, and I will attempt first of all to define it. It should first be pointed out that equilibrium is different from the so-called five senses, sight, hearing, smell, taste, and touch, all of which possess definite objects. Equilibrium means to be balanced. With respect to standing, it means one can remain standing in balance. From the standpoint of general physiology, equilibrium belongs to reflex or is a result of reflex. "Equilibrium reflex" is a more appropriate word than "equilibrium sense."

Equilibrium sense was also called labyrinthine sense and was considered to be closely related to the vestibulum and labyrinth. It was considered to be a sense which gave rise to the labyrinthine reflex and was inseparably connected to that reflex. I suspect that the word "equilibrium sense" was created when one assumed a labyrinthine receptor which evoked the labyrinthine reflex for the purpose of maintaining equilibrium. The sensation produced by this receptor was then assumed. However, a reflex is not necessary when one senses that one's body is in equilibrium. The reflex needs to be provoked when equilibrium is disrupted. When dysequilibrium is felt, the reflex is called for and equilibrium is restored. When body equilibrium is disrupted while standing or performing any other body movements, that is, when antigravity action is destroyed, body dysequilibrium is sensed and a reflex is provoked in order to restore equilibrium. Therefore, one should talk about "dysequilibrium sense" and "equilibrium reflex."

In the following, the question of body equilibrium in man, the objectivity of "equilibrium sense," and the labyrinthine physiology which gives rise to the "equilibrium sense" will be discussed along with a continuous examination of the expression "equilibrium sense."

14.2. Standing Posture and Equilibrium Reflex

The standing posture of man is quite an unstable posture. Numerous joints of the body assume almost one-point contact, and yet standing posture is maintained. Piling up the bones of the body so as to make the skeleton stand upright without motion is hardly possible. That the upright human body which contains these bones can perform delicate and complicated movements without falling by means of equilibrium action is quite amazing. The human body, which is unstable, can stand up against gravity and perform movements skillfully without exerting particular effort for equilibration. This mechanism was examined by our predecessors in this field and was found to be integrated reflexes which skillfully control the skeletal muscles of the entire body. The expression "equilibrium reflex" is not used very frequently in the existing literature, but I use it as I believe it more important than the expression "equilibrium sense." In the following, the concept of equilibrium reflex will be delineated.

Plants are supported by roots, which keep them erect on the ground. Their standing against gravity is easily explainable by physics. Humans and other animals do not counteract gravity by means of roots but stand using their own force or muscles. The standing posture of animals is quite unstable with respect to the center of gravity. Animals move by making use of this instability. Take a tortoise, for example. In this animal, the center of gravity is located lower than in most other animals, giving great stability to it. However, when it walks, it stretches its four limbs as illustrated in Figure 14.1, thereby raising its center or gravity to assume an unstable posture, which makes it possible to walk. In other words, the muscles are tensed in an antigravity fashion, and the center of gravity is raised. The instability thus created is utilized for walking. Walking is performed with a combination of the physical effect of falling and muscle movement which counteracts gravity. Consequently, the higher the center of gravity, the narrower the area to support the body and the more unstable the body shape, and the greater is the ability to move. This is illustrated in Figure 4.5 in Chapter 4 with respect to the tortoise, rabbit, dog, monkey, and human.

There are expressions concerning posture as immovable position, "motionless as a stone," etc. However, classical physiology proved that there is no such standing posture which is truly immovable or absolutely motionless. No matter how motionless a person tries to be, he assumes his standing position swaying constantly. The famous experiment with a

FIG. 14.1

cephalogram illustrates this well. In this experiment (Basler, 1929), the examinee wears a helmet with a needle on top of it that touches paper covered with soot so that the movement of the needle can be recorded. The cephalogram thus obtained shows that the standing posture, which was considered to be static, is actually maintained by incessant swaying (cf. Figs. 4.2 and 4.3 in Chapter 4). This holds true with normal healthy individuals and is different from the marked swaying which is clinically known as Romberg's phenomenon. It should be noted that the standing posture is maintained by swaying which can be objectively ascertained on cephalograms, although it is not as marked as in Romberg's phenomenon. A man is capable of standing because of incessant change in the tension of skeletal muscles. Recently, I conducted a fundamental experiment, in which an examinee stands with a water-filled glass on his head. Observations on some examinees, who were simply asked to stand quietly, showed that the water in the glass did not remain motionless but swayed enough to create waves. This of course was seen in normal healthy individuals.

The posture of standing is basic in humans and animals. It is maintained by continuous, unstable swaying. In order to maintain this standing posture, it is necessary to have the tension of strong skeletal muscles which counteract gravity. In other words, antigravity tension of skeletal muscles is indispensable. Studies have shown that this tension is brought about by way of muscular reflexes mediated by various receptors, namely, exteroceptors in the skin, proprioceptors, the vestibulum, labyrinth, and visual organs. The muscular reflexes mediated by these receptors maintain the standing posture, which is not immovable as a stone but is incessantly swaying as mentioned above. The concept of equilibrium is defined clearly in this connection. Muscular tension is adjusted subconsciously by each receptor responding delicately to the falling movement in which the center of gravity is attained lower and lower according to gravity. This adjustment is objectively observed as cephalographic fluctuation or the movement of the water contained in a glass on top of an examinee's head. This is what I call equilibrium and the equilibrium reflex.

The equilibrium reflex is usually restricted to the labyrinth but should be interpreted in a wider context. In the tabes dorsalis, in which the proprioceptors are impaired, there is marked equilibrium disturbance; a healthy individual produces marked cephalographic fluctuation if his eyes are closed; Romberg's phenomenon becomes more pronounced when the eyes are closed. The proprioceptor and the visual organ are involved in body equilibrium by mediating the equilibrium reflex before the sensation is consciously felt by the cerebral cortex. It is needless to say that equilibrium is disturbed when the labyrinth is impaired: this is well represented in various experiments with animals whose labyrinthine function was destroyed as well as in clinical cases. Although mention was made in the above only to the receptor, the involvement of the central nervous system, the cerebellum in particular, is also important.

What should be noted is that these receptors are not involved with body equilibrium independently, but that they are all closely related. This relationship can be clearly seen between the visual organ and the labyrinth. I believe that the labyrinth plays the subordinate role of supporting the optic labyrinthine reflex, which will be exemplified later. The relationship between the proprioceptor and the labyrinth has not yet been clearly delineated. However, careful analysis of the goniometric examination, a frequently employed function test, reveals that the labyrinth has an important relationship with the equilibrium reflex via deep sensation. This is a new area which needs to be developed in the future. I want to emphasize that the labyrinth performs the equilibrium reflex in close relationship with other receptors in a way that has hitherto been unknown: the labyrinthine reflex has to be considered as part of the equilibrium reflex.

14.3. Evaluation of "Equilibrium Sense"

It is considered that labyrinthine stimulation causes one to have sensation, the content of which is expressed in certain ways. It is stated that one feels equilibrium sense, position sense, motor sense, or labyrinthine sense via the labyrinthine receptor. Although one talks of the way he feels as the state of being in equilibrium, the content of such a sense is quite elusive. The equilibrium reflex can be considered to be brought about when one feels dysequilibrium, which is a more logical way of thinking. However, when one is talking about this feeling of dysequilibrium he is referring to the feeling of inclination or dislocation of the body trunk or head. This is a sense of position or movement and should be so designated rather than as equilibrium sense. It should be noted here that no one without labyrinthine function, when his eyes are closed, complains of being unable to know the position of his body trunk or head or to have no sense of motion when moving. Neither has it been reported that such senses are particularly dull in those with no labyrinthine function. What is specific to them is not the absence of such senses, but the fact that they are incapable of equilibration and that, particularly when the eyes are covered, marked equilibrium disturbance or lack of labyrinthine equilibrium reflex is produced. This is a very important point.

The labyrinth is a remarkable receptor with an elaborate structure. The cupula of the semicircular canal and the otolith of the vestibulum are just as elaborate as the auditory or visual organs. Its central pathway, however, is different from that of audition, vision, or the other major senses. The vestibular nerve is unique: it enters the vestibular nucleus at the brain stem, where it forms reflex pathways to the ocular, skeletal, and smooth muscles throughout the body, but it has no central pathway extending to the cortex. The vestibular nerve lacks a cortical area or ascending fibers, unlike the cochlear nerve or optic nerve that send some fibers to the reflex arc while the majority of them ascend to the cortex to form either the auditory or visual area. Figure 14.2 shows the central pathway of the vestibular nerve. The most proximal extent of the ascending fibers is the oculomotor nerve nucleus, beyond which no fibers are sent out. In contrast, the cochlear nerve imparts some fibers to the reflex arc, but most fibers ascend to the cortex to form the auditory area as illustrated in Figure 14.3. This is the most important basis of labyrinthine physiology. The visual or auditory organ receives stimulation, which then reaches the cerebral cortex to produce the senses of vision and audition in man. Meanwhile changes in muscular tension are brought about throughout the body by way of the reflex arc, which receives fibers from the brain stem or above. The stimulus received by the labyrinth, in contrast, only produces muscle reflexes by way of the reflex arc and gives rise to no primary sensation, which is the physiological viewpoint

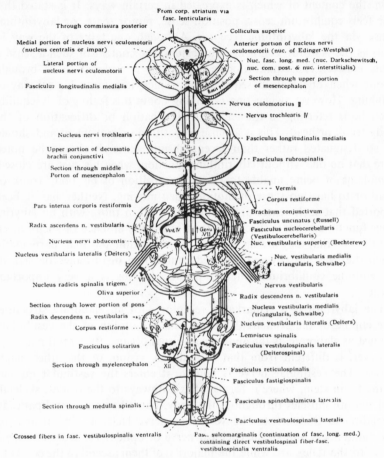

FIG. 14.2.   Vestibular system. The proximal extent of the ascending
fibers is the nervus oculomotorius, beyond which no fibers
are sent out.

derived from the anatomical facts that the central pathway does not
extend beyond the brain stem and that there is no corresponding cortical
area. The vestibular nervous system should not be treated as a sensory
system but be considered as a reflex system for equilibrium regulation.

What are then the senses of equilibrium, labyrinth, position, and
motion which are usually mentioned? These are not the primary sensations produced directly by labyrinthine stimulation. The stimulus received by the labyrinth changes and adjusts, by way of the reflex arc,
the tension of skeletal muscles of the body, thereby maintaining equilib-

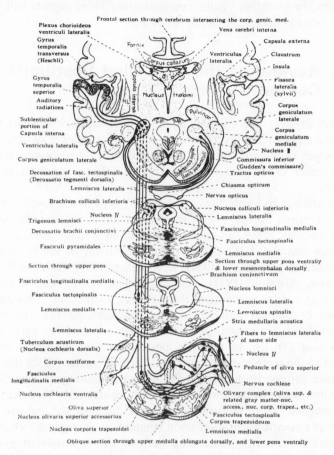

Frontal section through cerebrum intersecting the corp. genic. med.

Plexus chorioideus ventriculi lateralis
Gyrus temporalis transversus (Heschli)
Gyrus temporalis superior
Auditory radiations
Sublenticular portion of Capsula interna
Ventriculus lateralis
Corpus geniculatum laterale
Decussation of fasc. tectospinalis (Decussatio tegmenti dorsalis)
Lemniscus lateralis
Brachium colliculi inferioris
Nucleus IV
Trigonum lemnisci
Decussatio brachii conjunctivi
Fasciculi pyramidales
Section through upper pons
Fasciculus longitudinalis medialis
Fasciculus tectospinalis
Lemniscus medialis
Lemniscus lateralis
Tuberculum acusticum (Nucleus cochlearis dorsalis)
Corpus restiforme
Fasciculus longitudinalis medialis
Nucleus cochlearis ventralis
Oliva superior
Nucleus olivaris superior accessorius
Nucleus corporis trapezoidei

Vena cerebri interna
Capsula externa
Claustrum
Insula
Fissura lateralis (sylvii)
Corpus geniculatum laterale
Corpus geniculatum mediale
Nucleus II
Commissura inferior (Gudden's commissure)
Tractus opticus
Chiasma opticum
Nervus opticus
Nucleus colliculi inferioris
Lemniscus lateralis
Fasciculus longitudinalis medialis
Fasciculus tectospinalis
Lemniscus medialis
Section through upper pons ventrally & lower mesencephalon dorsally
Brachium conjunctivum
Nucleus lemnisci
Lemniscus lateralis
Lemniscus spinalis
Stria medullaris acustica
Fibers to lemniscus lateralis of same side
Nucleus IV
Peduncle of oliva superior
Nervus cochleae
Olivary complex (oliva sup. & related gray matter-nuc. access., nuc. corp. trapez., etc.)
Fasciculus tectospinalis
Corpus trapezoideum
Lemniscus medialis

Ventriculus lateralis
Fornix
Corpus callosum
Nucleus thalami
Pulvinar
Col Sub

Oblique section through upper medulla oblongata dorsally, and lower pons ventrally

Fig. 14.3.   Auditory system. Most fibers ascend to the cortex to form the auditory area.

rium. In this way, the standing position is maintained, or equilibrium is maintained dynamically at each moment of movement, thus allowing performance of smooth motions. Such a labyrinthine muscular reflex for equilibrium adjustment throughout the body is integrated in the cortex and perceived as the sense of position, motion, or labyrinth by consciousness. Thus, it can be seen that these senses do not originate in the vestibular nerve but are derived secondarily via the change in muscular tension.

I have repeatedly stated that there is no cortical area for the vestibular nerve. However, there have been electrophysiological studies which report on the so-called vestibulocortical area. These reports will be reviewed with my comments in the following.

14.4. Review of the So-Called Vestibular Center

Histologically, no fibers ascending from the vestibular nucleus to the cortex have been identified. However, there are a number of areas in the cortex which have been reported as being the vestibular center; this vestibular center is reported to be located in different areas by different reporters.

In an attempt to locate the corticosensory area of the vestibular nerve in the cerebrum, Spiegel (1932) strychnized the cortex and obtained important findings. Since this experimentation, numerous follow-up studies have been conducted. Electrophysiological pursuit of the cortical area was conducted by Spiegel (1934), Kornmüller (1937), Gerebtzoff (1940), Motokawa, Aronson (1933), Penfield, Poljak (1932), and Kempinsky (1951). Figure 14.4 summarizes these studies. Gerebtzoff stimulated the labyrinth of the cat cerebral cortex and detected the response in the ectosylvius girus and posterior to the suprasylvius gyrus (Brodmann's area 21), i.e., in the area posterior to the auditory area and adjacent to the temporal extrapyramidal field. Foerster (1932) stimulated the fissura interparietalis electrically during neurosurgery. Penfield gave electrostimulation to the temporal lobe intraoperatively. They each called the stimulated area the vestibular center because the patients complained of a rotation sensation upon stimulation. Kempinsky electrically stimulated the anterior descending limb of the suprasylvian gyrus by

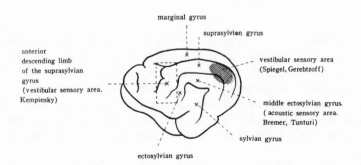

FIG. 14.4.   Corticosensory areas in cat cerebrum.

a method that he devised and described this area as the vestibular sensory area. He claims that with his method, no excitation was detected in the areas reported to be the center by Spiegel, Gerebtzoff, or Aronson. This short summary shows how diverse the opinions and methods are for the location of the vestibular center. The cochlear nerve, in contrast, has been worked out anatomically with respect to its center and reflex pathway, whereas only the reflex pathway has been delineated with respect to the vestibular nerve. The center of the vestibular nerve has been studied by various complicated methods, which resulted in diverse localities being designated as the vestibular center, as described earlier. What is noteworthy is that these possible center areas are located away from the pyramidal and sensory areas, e.g., in the extrapyramidal area in Penfield's field 22. It is to be expected that the vestibular nerve is related to the *adversive Felder* of the extrapyramidal tract, electrostimulation to which induces *adversive Bewegung*. This is to be expected in an evolutionary sense. However, I do not think it necessary to term it the vestibular sensory center as if the vestibular nerve sense reached there. What should be reconsidered is the habit of calling the area in the cerebral cortex which is excited on certain stimulation the sensory center for that nerve. The extrapyramidal area receives fibers from the high-order reflex arcs. Many reports of the so-called vestibular center may be of the point through which the secondary fibers of these high-order reflex arcs pass.

14.5. Vertigo and Motor Ataxia Produced by Labyrinthine Stimulation

The following describes how Foerster came to think that the fissura interparietalis, stimulation to which gave rise to the rotation sensation, was the vestibular center (Fig. 14.5). This interpretation is the natural outcome of the fixed idea that rotation sensation originates in the semicircular canal. However, there are various questions to be answered in regard to this thinking. In order to understand this issue, one needs to know about the sense of rotation, rotatory vertigo, and motor ataxia brought about via the labyrinth and about the complexity of the labyrinthine function. Thus I would like to point out that the labyrinth is an important reflex organ in charge of body equilibrium and that it easily brings about vertigo and motor ataxia with unusually strong stimulation and disrupts equilibrium instead of maintaining it.

As repeatedly stated, the labyrinth delicately responds to fine changes in rotation, inclination, or air pressure in order to perform muscular reflexes so as to adjust muscle tone relative to the stimulation, thereby

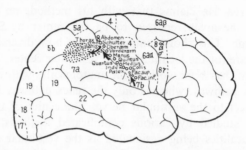

FIG. 14.5. Fissura interparietalis, which Foerster thought of as the vestibular center. Arrows indicate reflex pathway from fissura interparietalis to vestibularis.

maintaining body balance. If the head is tilted, a reflex is produced in the ocular muscles so that the eyeballs will resume the normal position as seen in *kompensatorische Augenstellung*. With respect to the skeletal muscles, the labyrinthine reflex is produced in the lower limbs when one is placed on a goniometer, for example, in order to retain the standing posture. Tokizane, Morimoto, Honjo, and we have reported that such changes in the muscle tone as are not apparently obvious can be identifiable as the labyrinthine equilibrium reflex by means of electromyography. I maintain that this reflex is in the *koordiniertes Stadium* (stage of coordination) of the labyrinthine reflex and that the labyrinthine equilibrium reflex is in this stage. The reflex in the coordinated stage is the essence of labyrinthine function. When this reflex is produced, there is no sensation of rotatory vertigo. However, when receiving strong stimulation which one is not accustomed to, the labyrinth produces vertigo and motor ataxia and disrupts equilibrium. I called this *geschlagenes Stadium* (stage of disruption). This is clearly represented in Bárány's rotation examination, an equilibrium function test, in which the examinee develops severe nystagmus after rotation and falls to the floor because he is unable to maintain the standing position; i.e., severe motor ataxia is produced. The direction of postrotatory sensation is opposite to that of the original rotation. This is accompanied by a marked rotatory vertigo. Foerster stimulated the fissura interparietalis (area 5), which resulted in a rotation sensation. On account of this, he called this area the vestibular center. I do not consider this rotation sensation to be the labyrinthine sensation. What is then the sensation one feels during rotation and rotatory vertigo after rotation?

The lymph flow theory presupposes the presence of movement of the cupula, which has been substantiated by Steinhausen (1925) who used a

FIG. 14.6.   Movement of cupula in the experiment on a hecht by
Steinhausen (1925).

hecht to prove it (Fig. 14.6). He claimed that rotation of constant velocity is not sensed; that angular accelerations of one degree per second
or less are not sensed; and that the cupula is pointed in a certain direction after the rotation is stopped, thereby producing the sensation of
rotation or rotatory vertigo in spite of the fact that the body itself has
ceased to rotate. Various rotation examinations, however, show that the
sensation one feels during rotation is not unequivocal: one person feels
as if he were rotating in the direction of rotation, another feels the opposite, and yet another feels no sensation of rotation. After rotation,
however, the sensation of rotating in the direction opposite to the rotation develops uniformly. What is unequivocal in rotation examination is
that the postrotatory vertigo is in the opposite direction to the rotation.
After lengthy rotation, however, second and third phases are observed,
and when the direction of nystagmus changes, that of rotatory vertigo
also changes. I would like to stress here that the rotation sensation is
quite unclear. The rotation sensation is not peculiar to the labyrinth.
As I reported elsewhere (Fukuda, 1957), clear rotatory vertigo can be
induced optically. Also, there are reports that an auditory or tactile
sensation can give rise to rotation sensation. Therefore, this area cannot
be called the vestibular center without incongruency. It should be called
the center of rotation sensation. The interparietal fissure has not been
shown to be closely connected with the vestibular nerve anatomically
or physiologically.

Here I will discuss the undeniable rotation sensation which always ac-
companies the nystagmus provoked either by Bárány's rotation or cold
water infusion into the external auditory meatus. This phenomenon is
simplistically interpreted to be produced by the lymph flow brought
about by the connection due to rotation or cold stimulation, giving rise
to the sense of rotation together with the ocular reflex. However, on
the basis of the anatomy of the vestibular nervous pathway, such sensation
cannot be considered to be produced primarily. With such strong stimula-

tion, the reflex of what I call the disrupted stage is provoked and body equilibrium is utterly disturbed. The feature of this dysequilibrium is that muscular tension becomes markedly deviated toward one direction; the direction of nystagmus, or of Romberg's phenomenon, is uniform. The sense of rotation has the same direction (slow phase of nystagmus or the direction of falling in Romberg's phenomenon). In this way, the sense of rotation is produced when the muscles of the body are abnormally tensed by an excessively stimulated labyrinth. This stimulation is transmitted via the deep sensory pathway to the cortex, where the tension is integrated to give rise to the sense of rotation.

In cupulometry, which recently became popular in northern Europe, slow rotation is employed because Bárány's rotation stimulation is considered to be too strong and nonphysiological. Even with this mild stimulation, a sense of rotation does not fail to appear after rotation. In cupulometry, the emphasis is on measuring labyrinthine sense or equilibrium sense. This fact raises an interesting question when elucidating labyrinthine physiology. Cupulometry will be introduced and evaluated briefly in the following section.

14.6. Introduction and Evaluation of Cupulometry

Bárány's 10 rotations in 20 seconds were criticized as excessive labyrinthine stimulation by researchers who advocated cupulometry. Figure 14.7 shows the device developed for cupulometric examination. A chair is rotated at a slow angular acceleration of $0.3°/sec^2$, then stopped when a certain constant angular velocity is attained and the after-nystagmus as well as the sensation of rotation are measured. This after-sensation is felt by the examinee immediately after the cessation of rotation and in the opposite direction to the rotation itself. In the theory of cupulometry this sensation is considered to be physiological for the labyrinth; this after-sensation originates in the labyrinth and is treated as the labyrinthine or equilibrium sensation.

According to the lymph flow theory, the flow of endolymph produced by rotation continues to be present even after the cessation of rotation, thereby causing distortion of the cupula (Fig. 14.6), which then gives rise to after-nystagmus as well as the sensation of rotation. Suppose that the cupula is elastic and is completely blocking the lymph space; then the intensity of the rotation sensation brought about after the cessation of rotation is proportional to the distortion of the cupula, and the duration of this sensation coincides with the time required for the

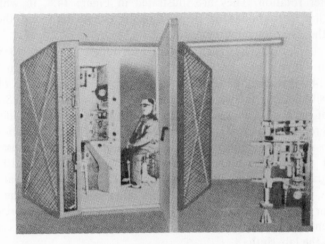

FIG. 14.7.   Cupulometric examination. Reproduced from Wit (1953).

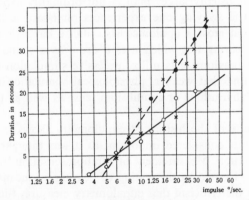

FIG. 14.8.   Cupulogram in normal subject. Solid line shows sensation
cupulogram; dotted line, nystagmus cupulogram. Reproduced from Wit (1953).

cupula to resume its initial position. An angular acceleration of $0.3°/sec^2$
has been employed because no inertia of endolymph is produced by
acceleration slower than that, and any desired rotation velocity can be
used without causing a rotation sensation when the velocity is increased.
When rotation is suddenly stopped, after-nystagmus and rotation sensation comparable to the rotation velocity are obtained. In other words,
nystagmus or rotation sensation is not produced during rotation but

only after rotation. They are illustrated in Figure 14.8, in which the dotted line corresponds to the nystagmus cupulogram and the solid line to the sensation cupulogram.

This cupulometry has been applied with different results to various cases of labyrinthine impairment, motion sickness, and brain tumor. Figure 14.9 is a study of seasickness. ISS represents the group of those who easily become seasick, and the slope of the line is steep. NSS is the group of those who do not get seasick, and the slope of the line is not steep. S represents normal people, and the slope is between the above two. The sensation cupulogram, indicator of the examinee's after-sensation, is selected rather than the nystagmus cupulogram. Figure 14.10 shows cupulograms taken after Bárány's rotation. The solid line is that before rotation, the interrupted line one hour after rotation, and the dotted line four days afterwards. The slope of the cupulogram decreased as time went by, which leads to a curious conclusion, that Bárány's rotation impairs the labyrinth.

In cupulometry, careful attention and consideration are given to stimulation. Bárány's rotation utilizes an angular velocity of 180°/sec, which is unusually large: man is not ordinarily exposed to such stimulation. The swaying one experiences on board a ship has a velocity of 60°/sec at most. Cupulometrists criticize Bárány's test by comparing it to the hearing test which uses a drum as a sound source. They consider that previous tests utilize excessive stimulation which is not physiological, and that weaker, physiological stimulation should be given in study of labyrinthine or equilibrium sense as well as the labyrinthine reflex.

The study of the history of cupulometry shows that the sensation of rotation was considered important as it was thought to be the labyrinthine sensation. Measurement of rotation sensation by changing rotation velocity is likened to the testing of hearing by means of various tuning forks. To the criticism that cupulometry measures function by analyzing the subjective sense of rotation, cupulometrists say that auditory examination also utilizes subjective judgment. The sensation of rotation is considered very important, which is well illustrated in a paper on seasickness by van Egmond (1953). He measured only the sense of rotation, i.e., sensation cupulograms, and did not refer to the nystagmus cupulogram after rotation at all.

Cupulometrists treat the vestibular nerve exclusively as a sensory nerve and the labyrinth exclusively as a sensory organ, a very clearcut viewpoint. The labyrinthine reflex is considered to be secondary and accessory. This situation is likened to the pupillary reflex in relation to visual stimulation.

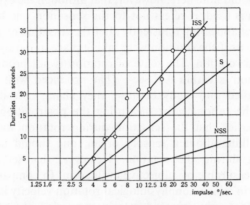

FIG. 14.9. Cupulogram in study of seasickness. ISS, NSS, and S represent a subject who becomes seasick, one who does not, and a normal person, respectively. Cupulogram of ISS shows a steep slope, while that of NSS has a weak slope. S presents an intermediate slope. Reproduced from Wit (1953).

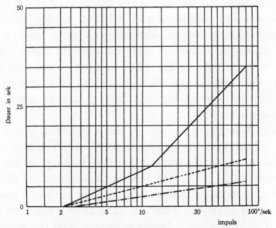

FIG. 14.10. Cupulogram before and after Bárány's rotation. Solid line, before rotation; interrupted line, one hour after rotation; dotted line, four days after rotation. Reproduced from Rossberg (1955).

One should note here, however, that although the expression of rotation sensation is used, it refers not to the sensation one has during rotation but to that appearing transiently after the cessation of rotation. This is experienced as if the rotation were taking place in the opposite direction to the initial rotation. The sense of rotation is not produced if the angular acceleration is within $0.3°/sec^2$. That produced with faster acceleration is not called into study because, they claim, it is difficult to measure. This is the rationale for measuring the postrotation sensation. I mentioned earlier that the sensation during rotation was not constant. If one is talking about a sensory organ, one should be dealing with the sensation during the rotation. However, this sensation is ill-defined and unmeasurable by subjective evaluation. In comparison with Bárány's rotation, in which nonphysiological rotation velocity is employed, cupulometry adopts physiological velocity, and thus the after-nystagmus and after-sensation are considered to be a physiological sensation and reflex. After the body of an examinee has ceased to rotate, he feels as if his body were rotating in the opposite direction. Such a sensation is considered physiological by the theory of cupulometry, a very bizarre concept. One feels as if one were rotating in the opposite direction when one actually is not; such a sensation is nothing but illusion, and is the very definition of the sense of vertigo. Nevertheless it is claimed that Bárány's rotation is too strong and causes vertigo, while the cupulometric stimulation is physiological and produces a physiological sensation but no vertigo. This seems an astounding claim to me. One is reminded here of the words "equilibrium" and "equilibrium sense," which I described as unclear concepts earlier. The sensation of rotation, which is the indicator in cupulometry, i.e., illusion, vertigo, or a sense of feeling as if being rotated in the opposite direction after rotation, seems to be part of the "equilibrium sense."

I have just referred to one of the after-sensations, a postrotatory sensation of rotation, using cupulometry as an example. In the so-called equilibrium sense, definite sensation is lacking, as stated earlier. What is definitely felt is this after-sensation—vertigo or illusion. This is not restricted to rotation and can be applied to linear movement. Take an elevator, for example. What one feels clearly while on an elevator is the sense of sinking when it begins to ascend, and the sense of floating when stopping. These two sensations are clearly felt. Our experience shows that during ascension or descending one is hardly aware of the linear movement if one has one's eyes closed. When one analyzes the equilibrium sense, one finds that it is not felt during the movement but before or after rotation or linear movement as a definite illusion or vertigo; this is a truly strange situation. The sense of equilibrium is a quite ambiguous sense, but on analysis one comes to such an illusion or vertigo.

The theory of cupulometry in essence boils down to the question of the sense of rotation. It has been treated as the physiological sensation of the labyrinth, which is nothing but vertigo. Therefore, a strange conclusion is obtained, that the labyrinth is a vertigo-sensing organ. Certainly the labyrinth brings about vertigo and dysequilibrium depending on the stimulus. However, this is just one aspect of the labyrinthine function. The main function of the labyrinth is to control equilibration, which has been shown by numerous animal experiments and clinical examples. Then why not emphasize the equilibrium function when studying labyrinthine function? Earlier I stated that the labyrinthine function is purely reflexive, a conclusion reached only after faithfully observing the reflexive function of the labyrinth. Analysis of the sensation experienced during linear or rotatory movement reveals that it is a very unstable and unclear one. Study of persons with complete or incomplete labyrinthine impairment leads to the conclusion that the labyrinth cannot be considered to be a sensory organ. What is certain is that a sense of movement or vertigo in the opposite direction to the original motion is present. It is also important that this vertigo is always accompanied by motor ataxia and that it is reduced or disappears following repeated stimulation, i.e., training. These findings led me to abandon the idea of regarding the labyrinth in the same vein as the visual or auditory organs. I regard the labyrinth as a purely reflexive organ and classify the phases of reflex into those of coordination and of disruption. I gave a lengthy exposition of cupulometry, because I wanted to observe nature faithfully without fixed ideas. If one studies rotation sensation with the assumption that the labyrinth is a sensory organ dealing with the equilibrium sense or with the lymph flow theory, thereby entertaining fixed ideas and simplifying labyrinthine physiology, one ends up concluding that illusion and vertigo are the physiological labyrinthine sense or equilibrium sense.

In comparison with Bárány's rotation examination, cupulometry to me is one step forward and two steps backward; the former because it paid attention to rotation, and the latter because it regarded vertigo or illusion, postrotatory sensation of rotation, as the physiological labyrinthine sense.

In cupulometry, great attention is paid to rotation velocity; very slow rotation stimulation, which everybody considers physiological, is applied to the labyrinth. Even after this weak stimulation, the labyrinth gives rise to vertigo (sense of rotation) and motor ataxia (after-nystagmus), which I call the phase of disruption. Is the labyrinth such a friable organ that a carefully chosen weak rotation stimulation, which is considered appropriate, produces equilibrium ataxia? This question has to be solved.

I consider that the problem lies in this rotation stimulation, which will be discussed in the following section.

14.7. Reevaluation of Labyrinthine Stimulation

Various methods of stimulation have been tried in labyrinthine physiology and have been reasonably divided into appropriate and inappropriate stimulation. To the former belong linear movement, tilting, and rotation, whereas to the latter belong temperature, electricity, and various types of inner ear damage from ear disease. This classification appears consistent with the structure of the inner ear.

It is important to see whether these so-called appropriate stimuli had a proper place in evolutionary history. In laboratories for labyrinthine stimulation, we place animals on a rotating table for rotation, on a goniometer for tilting, and on an elevator for linear movement without much thinking. These experimental manipulations are in fact more appropriate for the structure of the labyrinth in comparison with temperature, electricity, or destruction experiments. Cupulometrists regard 10 turns in 20 seconds nonphysiological and employ angular acceleration of $0.3°/sec^2$ or less to limit angular velocity within $90°/sec$, which they consider more appropriate. What is important is to evaluate first whether such changes in position and movement actually exist in the life of animals, mammals in particular. As there have been no questions or issues raised with regard to this problem, I would like to point out the following.

All the above-cited stimuli are passive changes in position or movement relative to the body of animals. The table, board, or floor on which animals are placed is moved passively. These changes mean to animals that the ground on which they are standing is suddenly tilted, rotated, or moved up or down. Have they ever received such stimulation before they are brought to the laboratory? Although cupulometrists criticize Bárány's rotation as nonphysiological and employ what they call physiological rotation stimulation, such stimulation of the ground's being turned around has never been experienced by animals and, in this sense, is nonphysiological.

The point I would like to make is not that animals never experience the stimuli of rotation, tilting, or going up and down, but that such stimulation is experienced by the animals when they themselves move actively. Labyrinth destruction experiments in animals shows that the first development is inability to maintain normal posture or perform normal movement—impairment in active movement that is termed equilibrium disturbance or motor ataxia. This applies to human beings with impaired labyrinthine function, who complain of feeling unstable and are incapable of standing or walking, i.e., the definition of motor ataxia.

Summarizing the above, I would like to suggest that the muscular reflex produced by imposing passive changes in position or movement on animals or human beings cannot simply be equated with the labyrinthine reflex. Such passive stimulation does not exist in animal life or its evolution. The essence of the labyrinthine reflex should be sought not in the reflex provoked by such passive changes but in situations where the labyrinth responds to active movement accompanied by changes in position or movement in maintaining equilibrium or performing movements.

I would like to analyze labyrinthine physiology on the basis of the above concept. When viewing the evolutionary history of mammals up to the time monkeys evolved, one hardly finds passive changes in position or movement inflicted on their bodies. The only conceivable passive movement in their life in trees would be the swinging of the trees: otherwise they move actively. As they are equipped with an elaborate vestibular organ, their labyrinthine function is considered to be primarily involved in active movement. The tilting or rotation of the ground on which animals stand is an extraordinary stimulation for them. It is to be noted that a new nuance has been added with the advent of human beings. The invention of vehicles has added a new factor of passive stimulation to the labyrinth. This situation will become clear if one thinks of the invention of palanquins, locomotives, steamships, and airplanes. The cause of motion sickness lies in this passive movement which human bodies experienced for the first time in their history. Humans have not been adapted to passive stimulation, which produces vertigo and motion sickness. My interpretation is that the labyrinth could not cope with the passive changes in position or movement from which motor ataxia results.

I should also point out that even with active movement, vertigo and motor ataxia can develop if performed excessively. One often sees children spinning around while playing and ultimately falling to the ground dizzy and incapable of standing, i.e., developing motor ataxia. On careful analysis, one finds that after-nystagmus develops after the cessation of rotation. This is similar to the development of optic vertigo, the condition in which excessive optic stimulation from rotation is imposed on the visual organ and produces vertigo and motor ataxia. This is simply a disturbance of body equilibrium caused by any receptor receiving excessive stimulation. Judging from the life and evolution of animals, the essential role of the labyrinth is in supporting active movement and allowing the smooth pursuit thereof. Therefore, when passive rotation is imposed on such a system whose primary target is in active movement, the after-sensation and rotatory vertigo of transient motor ataxia are easily produced, even with what cupulometrists call physiological rotation.

It can be summarized that passive movement, be it linear or rotatory, is not present in the life and history of animals. All their movements are active, and can be separated into linear and rotatory. To the former, the vestibulum responds swiftly moment by moment, and to the latter, the semicircular canal, in order to control muscular tension so that active movement can be performed. From this standpoint, animals do not have to sense or be aware of the position or movement of their own bodies while performing active movements. In other words, the necessity of the labyrinth's being a sensory organ is not very great. What is required most of the labyrinth is adjustment of muscular tension and performance of smooth movement in swift reponse to active movement of the body such as rotation, linear movement, or tilting—the moment-to-moment equilibrium reflex. Based on this concept, I claim that the labyrinth is not an equilibrium sensory organ but an equilibrium reflex organ with a projection and a reflex pathway. One may argue that as far as the otolith and cupula are concerned, dislocation of the stones or distortion of the cupula can be produced similarly, whether the movement is passive or active. However, the muscular tension completely differs between that produced by active movement and that by passive movement. Güttich (1920) and I (Fukuda, 1957) have reported that opposite, symmetric muscular tension was often produced. This example emphasizes the importance of evaluating the otolith or cupula not as something confined to the inner ear but as part of the whole system involved in controlling body equilibrium.

14.8. Criticism of "Equilibrium Sense"

When thinking of sensory organs in general, one finds that in the early stages of evolution they take the form of a receptor-reflex center-effector. This applies to the visual organ and organs of the five senses. Animals of lower classes lead an exclusively reflex life. As evolution progressed, the cerebral cortex was formed, and the sensory area was created. With this development, animals came to entertain the perception of the five senses. These senses are classified into perception (*Wahrnehmung*), sense (*Sinn*), and sensation (*Empfindung*) according to their nature. Visual, auditory, and the rest of the so-called five senses have been worked out anatomically so that the pathway from the receptor to the sensory sensor has been delineated. Perception, sense, and sensation have thus been substantiated. This applies in terms of brain anatomy to animals of higher class in which the cerebral cortex has been formed, namely, to mammals. In lower animals, the receptor is just an organ giving rise to the appropriate reflex. Although the receptor may be termed a sensory organ in higher animals, it produces reflexes subconsciously besides entertaining senses. This is illustrated also in man with respect to the five senses.

The labyrinth, however, is unique and different from the five senses. As stated already, anatomically the labyrinth lacks the ascending central pathway projecting from the brain stem. It does not possess a definite cortical area. What is definite is the presence of the receptor, the vestibular nerve, and the vestibular nucleus or reflex center. In other words, the feature of a sensory organ is absent and only the reflex mechanism is known. The structure of the receptor closely resembles that of a sensory organ and it is located just adjacent to the cochlea. For these reasons, many people regard the labyrinth as a sensory organ, look for its sensory center, and even devise various ways of sensory examinations for it. However, even when the labyrinth is treated as a sensory organ for position and motion, those who have little or no labyrinthine function do not ever complain that they cannot perceive position or motion or that they have become less able to do so than before. This is shown by the report of Grahe (1927) and numerous other reports. What is specific to these patients is not the absence or dulling of the equilibrium sense but motor ataxia or equilibrium disturbance appearing while performing active or passive movement. In other words, there is an absence or impairment of labyrinthine equilibrium reflex in supporting volitional movement.

There are two distinct ways of thinking in traditional labyrinthine physiology. One treats the labyrinth as a sensory organ; the other, as a purely reflex organ. From the former approach the word "equilibrium sense" was produced. Cupulometry is a typical product of this thinking. The latter idea of regarding the labyrinth as a purely reflexive organ is what I have come to. Reviewing the literature, one finds that Winkler (1921) advocated this idea. Magnus (1924), de Klijne and Versleegh (1930), and Rademaker (1931) share the same basic approach. There are two such ideas in labyrinthine physiology, although the presence of these two approaches is not clearly recognized by labyrinthine physiologists when they undertake research or experimentation. Because of this lack of recognition, researchers' opinions fluctuate between the two. This seems to be a peculiarity in labyrinthine physiology.

Among various textbooks on labyrinthine physiology, there are some in which the labyrinth is clearly treated as a sensory organ, although what they describe is mostly the labyrinthine reflex. Although the word "sense" is used in these books, it is restricted to the vertigo (illusion) produced via the labyrinth. However, it does not seem very scientific to treat the labyrinth as a sensory organ and to define the sense as equilibrium sense or labyrinthine sense. In many newer textbooks, the labyrinth is not dealt with as a sensory organ but as a reflexive organ in conjunction with various skeletal muscular reflexes such as the postural reflex including the neck reflex.

14.9. On Training

Based on these ideas, I will mention some of the training methods we have utilized. The history of evolution shows that when the environment changes, animals become equipped with an organ or organs appropriate for the change. This is exemplified in the evolution of the vestibular semicircular canal system from the vestibulum in lower animals. Animals equipped with a system of organs that seem to have completed evolution can develop a function(s) appropriate for new stimulations imposed by environmental change. This will be illustrated using our experiments on equilibrium function. When animals experience swinging or rotatory movements, they initially lose equilibrium and are unable to maintain the normal posture, because they are faced with passive change in position they have never experienced previously. However, as they receive such stimulation day by day, they become able to respond smoothly to this passive change in position and acquire the capacity to assume normal posture on the swing or rotation table. For example, chickens, which initially often fall off a swing without being able to adapt to swinging, acquire the equilibrium function through daily training to respond to the passive motion of swinging, thereby assuming a posture symmetric to that taken earlier (cf. Appendix A). This is a reflex movement induced by labyrinthine stimulation; it cannot be produced or is lost when the hyperstriatum is destroyed, and it can be said that it is a new equilibrium function created over the area from the vestibulum and labyrinth system to the cortex. The pre- and post-training postures are completely symmetric, although the stimulus the swing gives to the labyrinth stays exactly the same. The vestibulum and labyrinth may be a receptor organ which functions mechanically, but the equilibrium reflex manifests itself markedly differently, as in this result in chickens. What is important is that the vestibulum and labyrinth are capable of adapting to new environments and new vestibular stimuli. One tends to analyze the system purely physically in terms of otolith dislocation and cupula distortion and to retain the same approach when dealing with the higher center including the cortex. Such a rigid approach should not be employed when dealing with the structures situated more proximal than the peripheral organs. The example with chickens has shown that new adaptation has ensued in the vestibulum and labyrinth when a new stimulation was repeatedly given.

The same holds for man. Be it passive or active, repeated rotation stimulation results in marked reduction in after-nystagmus. In addition, the postrotatory sensation of rotation, regarded as important in cupulometry, is also markedly decreased or disappears.

We regard training as important for labyrinthine or equilibrium physiology, because the stimulation of passive movement has appeared only after the invention of vehicles, such as locomotives and steamships. Animal bodies are not used to such passive stimulation, and they are liable to vertigo and motor ataxia. In order to overcome this equilibrium disruption, various labyrinthine stimuli are repeatedly given to produce effective results in man, as shown in the example of chickens.

During the human developmental process, man receives various kinds of training in order to get used to these passive movements. Athletic facilities in parks, kindergartens, and elementary schools illustrate this clearly. Infants enjoy riding on wooden horses, which give a mild up and down movement, i.e., passive linear movement in the superoinferior direction. Man probably moved in a passive way for the first time riding on horseback, the prototype of vehicles. Slides, swinging poles, swings, and amusement park airplanes will all provide training in passive movement where linear and circular movements are combined. It is quite interesting that man is ready to take up various kinds of passive movement as he undergoes functional development beginning in infancy. This is not seen in the infancy of animals other than man. That humans who invented vehicles are ready to train themselves to adapt to passive movement from their youth so as to acquire equilibrium function seems to mean that man has known the importance of training the vestibulum and labyrinth.

As stated earlier, vertigo and motor ataxia can be minimized during passive movement by assuming the posture as if performing active movement. This is important from the standpoint of effective movement. When moving fast in a linear or circular fashion, as in ski jumping, horseback riding, or motorcycle riding, the body is moving in a passive way. However, movement becomes most effective by assuming the active posture as if actively performing a movement, such as jumping, riding, or rotating. This will enable man to handle such fast speeds as exceed the speed of his own running. The posture beautifully constitutes equilibrium function or motor function, which can be acquired through training of the labyrinth and vision.

14.10. The Labyrinth in Relation to Equilibrium Function

In conclusion, I would like to state that although the labyrinth regulates equilibrium, it is subordinate to other reflex systems. This can best be exemplified in nystagmus, which is considered important as it typically expresses labyrinthine function. The crab develops active optic nystagmus. When its vision is blocked, it does not develop rotatory nystagmus even with rotation stimulation. The crab possesses the statocyst but not the semicircular canal. This explains why it shows no nystagmus on rotation. What is important, however, is that active nystagmus is produced optically. This means that nystagmus is originally produced optically. After animals develop from the statocyst stage to possess the semicircular canal, the labyrinth participates in nystagmus development. In other words, the labyrinth plays a role in supporting the optic reflex (Figs. 14.11 and 14.12).

Numerous studies were conducted from the time of Grahe up to the development of cupulometry by treating the labyrinth as a sensory organ. However, none has successfully clarified the role of the labyrinth as a sensory organ for position and motion. This is quite different from the auditory or visual organ which can be treated and measured as a sensory organ and whose function can be expressed in numerical terms. As already stated, what is relatively clearly felt consciously as a sense induced by the vestibulum is vertigo. The postrotatory sensation of rotation in cupulometry and the sensation of floating or sinking when riding an elevator are vertigo. The word "equilibrium sense" refers to this vague sensation of mild vertigo.

All attempts to test the vestibulum using the sense of position or motion as the objective have failed. The present functional examinations all aim at the labyrinthine reflex. The labyrinth is treated purely as a reflex organ. Nystagmus, goniometry, and other tests all aim at the muscular reflex produced by labyrinthine stimulation. The word "equilibrium function examination" has replaced the word "labyrinthine function examination," which seems to coincide with the approach of considering the vestibulum as a reflex organ involved in equilibrium function. However, equilibrium function examination is complex. Standing, standing on one foot, and goniometry are genuine equilibrium examinations.

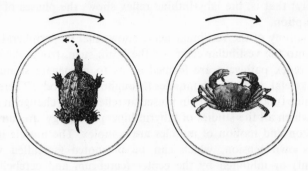

FIG. 14.11.   Rotatory stimulation with vision blocked. When rotated
with vision blocked, a tortoise develops nystagmus and
head and body deviations, whereas a crab, which has
no semicircular canal, develops neither nystagmus nor
body deviation.

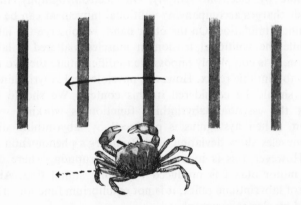

FIG. 14.12.   Crab in rotating cylinder. A crab develops nystagmus
and body deviation, even though it lacks the semicircular canal, when visual stimulation is given by objects
successively moving in front of its eyes. This shows
that nystagmus is produced originally by the optic
mechanism.

The examination of the nystagmus induced by rotation or temperature,
however, is not an equilibrium examination but a test of ataxia. This is
understandable if one thinks of vertigo and ataxia accompanying nystagmus. However, such an ataxia cannot be produced in those with no
labyrinthine function. As I have said, although the labyrinth regulates
equilibrium reflex, it disrupts equilibrium and motion when stimulated

excessively; that is, the labyrinthine reflex shows the phases of response and disruption.

The anatomy of the vestibular nerve shows that its peripheral pathway extends into the vestibular center of the brain stem, from which various complex reflex pathways are formed so as to propagate stimuli to the ocular, skeletal, and smooth muscles throughout the body. Therefore, the labyrinthine function appears in muscular reflexes and change in muscular tension which all the studies of labyrinthine physiology are directed to. The tension and motion of muscles are complex. The muscle itself possesses its own tension, which can be promoted (so-called volitional movement) or inhibited by the center (cerebrum and cerebellum) and modulated by various reflex systems including the labyrinth. Therefore, unlike vision or audition, it is almost impossible to differentiate the labyrinthine reflex from muscles. One has to regard the change in muscular tension produced by labyrinthine stimulation as the labyrinthine reflex. This change may be manifest, as in nystagmus, or latent to be only recognizable by electromyography. On electromyography, the interval of spike discharges accompanying volitional movement can be altered by labyrinthine stimulation. On the other hand, we observe the labyrinthine reflex while the volitional tension in muscles is altered by labyrinthine stimulation. It is completely impossible to differentiate the two to observe only the labyrinthine reflex. However, physiological labyrinthine function or reflex should be considered in this context. We should remember here that the essential labyrinthine function is working with active movement. When nystagmus is produced by labyrinthine stimulation, skeletal muscles show deviation, and Romberg's phenomenon is demonstrated. However, this is in the phase of disruption, where dysequilibrium or motor ataxia is produced by strong stimulation. Although it is a kind of labyrinthine reflex, it is not equilibrium function. This can be confirmed by electromyography.

Many scholars have been studying labyrinthine physiology. The approach of treating the labyrinth as a sensory organ ends up in a *cul-de-sac*, as I stated in relation to cupulometry. When studying the muscular reflex, there is a tendency to examine only the phase of disruption using excessive labyrinthine stimulation. As the essence of the labyrinthine reflex is the equilibrium reflex, labyrinthine function should be studied in this context. Equilibrium function and reflex are quite complex. The tension of muscles changes delicately second by second in order to maintain body equilibrium and allows smooth performance of motion. This muscular tension is modulated by various factors. The cerebrum, labyrinth, visual sense, touch sensation, deep sensation, and other reflex systems participate in the regulation of muscular tension. However, the change in muscular tension is expressed objectively only as change in the EMG spikes, regardless of the type of stimulation, be it labyrinthine or otherwise. In order to study labyrinthine function in terms of equilibrium reflex, one needs to study how the actually working equilibrium function changes with labyrinthine stimulation. As it is not possible just to take out the labyrinthine equilibrium reflex, one has to see how the overall equilibrium function is altered by labyrinthine stimulation or what defect is produced in the equilibrium function in those with impaired labyrinthine function. The labyrinthine reflex is observed only in the context of the overall equilibrium function, and this is the essence of the labyrinthine function.

Appendix A

Original Papers Published
in Western Journals

Reproduced with kind permission of Acta Otolaryngologica, Stockholm
(1–4, 6, and 7) and S. Karger AG, Basel (5).

STATIC AND KINETIC LABYRINTHINE REFLEX

Functional Development of Labyrinthine Function with Rotatory Training

TADASHI FUKUDA, MANABI HINOKI and TAKASHI TOKITA
Gifu, Japan

Department of Otolaryngology, Gifu Medical School
(Head: Prof. Tadashi Fukuda, M.D.)

Two blindfolded leghorns were rotated 100 times in 200 seconds to both directions every day for two weeks. Their labyrinthine function was evaluted by Bárány's rotation test before and after the repeated rotations.

It was observed that after the rotations the head during rotation turned in the direction of rotation after the normal deviation. This phenomenon was never found in animals before repeated rotations. The labyrinthine function which caused it has been named the "kinetic labyrinthine reflex". In contrast with this naming, the normal deviation during rotation, a hitherto well known labyrinthine reflex, has been called the "static labyrinthine reflex".

After repeated rotations the animal was much less ataxic during and after the test rotation: the animal showed a functional progress in equilibrating function through the repeated rotations, which were therefore named "training".

Postrotatory head-nystagmus was also much less marked in the trained animals.

INTRODUCTION

It has been proved that when animals receive rotatory stimulation every day, postrotatory nystagmus decreases. This hyponystagmus is understood from various standpoints: habit (1, 2, 3), fatigue (4), accomodation (5), damage (6), central effect (7), etc.

In addition to these points of view, Fukuda (8), one of the writers, has developed a new hypothesis to explain this hyponystagmus. During the period of postrotatory nystagmus the individual can be assumed to be in a state of transient artificial labyrinthine ataxia, a pathological state seen in normal subjects. As is well known from experiments in man, postrotatory nystagmus is accompanied by intense vertigo, inability to stand and a marked Romberg's phenomenon—a state of disturbed normal body equilibrium or ataxia. Thus it is his opinion that the decrease of postrotatory nystagmus in normal subjects, i.e. diminution of jerks and shortening of duration with repetition of rotation, should be considered as evidence that artificial ataxia is overcome and the labyrinthine equilibrating function is improved with that procedure.

The above view has been supported by the results of various experiments and tests carried out by the writers for more than fifteen years in Japan. Marked hyponystagmus, which means a lower degree of nystagmus in Bárány's rotation test than the normal average, has been observed in many dancers and athletes, who can be considered as imposing repeated rotatory movements upon themselves (9). It has also been found that after imposition of repeated passive or active rotations every day for one to three months many boys in a primary school showed marked decrease in postrotatory nystagmus accompanied by improved ability in various sports including gymnastics and also by improved epuilibrium as well as motor functions (10). Beacuse of these facts he calls the repeated rotations "training".

This report is concerned with decrease in postrotatory nystagmus of leghorns caused by repeated rotations. It differs, however, from many reports heretofore made on this subject in that a new concept of the equilibrating function is established in this report, which is based upon the observations of animals' posture during rotation: a new concept that a hitherto unknown new labyrinthine reflex of higher order is formed by means of repeated rotations.

EXPERIMENTAL

Experimental studies on decrease in postrotatory nystagmus due to repeated rotations have hitherto been made extensively by the writers in many animals as well as human subjects. In these experiments postrotatory eye-nystagmus was recorded on a subject having his head and body bound and fixed which was the routine procedure. Decrease in postrotatory head-nystagmus was studied in many hens and cocks, pigeons and tortoises, but in these cases the body was bound and fixed as well. It is apparent that under these conditions changes in vestibulo-spinal reflex (Deitero-spinal reflex) or postural changes cannot be observed. Thus an attempt was made to test postural changes by rotation of a leghorn standing freely on a rotating chair. However, it was found difficult for a free animal to maintain fixed position, and his active and passive movements made accurate observation of his postural changes difficult. Our final device was that a leghorn was made to alight freely on a perch prepared on a rotating chair, and this attempt was successful in fixing the position of the free animal during and after rotation. It took five years of effort to achieve this success.

Results of experiments in two representative leghorns

Two animals (white leghorns) were subjected to a certain amount of rotatory stimulation (repeated rotations) daily for two weeks. The animals were tested before and after the repeated rotations as well as on the third, seventh, and tenth days. The experimental data before and after the end of repeated rotations alone will be described. The abridged data concerning postural changes and decrease in post-

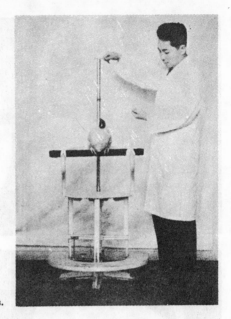

FIG. 1. The apparatus.

rotatory head-nystagmus which were tested on the intervening days showed a
gradual shift from the initial to the last data.

Test rotation.—The animal was blindfolded with a black band and settled on a
perch in the center of Bárány's chair (Fig. 1), and then subjected to 20 rotations
in 10 seconds (Bárány's rotation test). The animal's posture was observed during
and after the rotation and the jerks of postrotatory head-nystagmus were calculated.
The test rotation was done in both directions.

Repeated.—100 rotations in 200 seconds under the same conditions as the test
rotation in both directions.

Results.—The repeated rotation was extraordinarily strong for the animal and
it fell down frequently from the perch during the rotation in earlier days. Ataxia
was more severe when the rotation ended and the animal fell down each time.
But the ataxic behaviour became less marked as the rotation was repeated and
after two weeks the animal did not show marked ataxia during rotation. After the
rotation, however, the animal at times showed poor equilibrium, and scarcely
kept itself on the perch. Both animals did well without any abnormal behaviour
throughout the experiment.

*Results of the test before repeated rotations in the first animal following rotation
to the left.*—The animal's head and neck deviated about 90° to the right immediately
after beginning the rotation (Fig. 2). Then, the animal had a perrotatoric head-
nystagmus to the left which ceased 10 seconds after the beginning of the rotation.
The position of the head was kept deviated at 60°–90° to the right after that. That
is, not only the head and neck but also the body was deviated against the direction
of rotation, and the animal flapped its wings and barely kept itself on the perch
(Fig. 3). When the rotation ended, deviation of the posture shifted to the left of

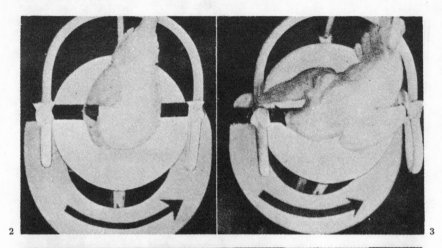

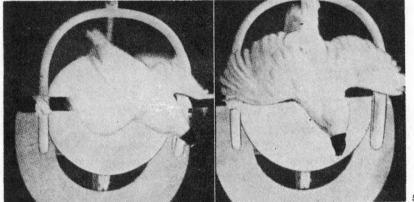

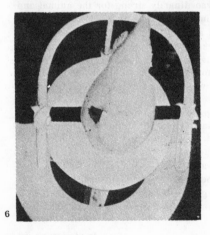

Before Training. (The direction of rotation is to left. The arrow indicates rotation and its direction.)

FIG. 2. Two seconds after the beginning of rotation. The head deviates in the opposite direction of rotation: the static labyrinthine reflex (normal deviation).

FIG. 3. Fourteen seconds after the beginning of rotation. Still, the static labyrinthine reflex (normal deviation).

FIG. 4. 0–2 seconds after the cessation of rotation to left.

FIG. 5. Twelve seconds after the cessation of rotation. Postrotatory head-nystagmus occurs to right.

FIG. 6. Twenty-five seconds after the cessation of rotation. The second phase of postrotatory head-nystagmus begins.

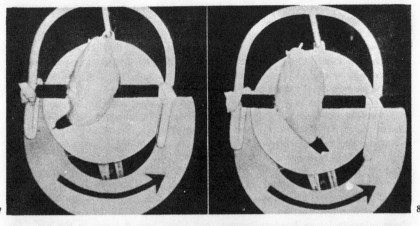

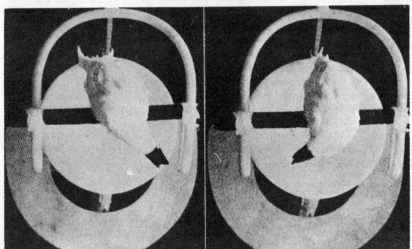

After Training.

Fig. 7. Two seconds after the beginning of rotation. The head deviates also in the opposite direction of rotation: the static labyrinthine reflex (normal deviation).

Fig. 8. Eight seconds after the beginning of rotation. The head turns in the same direction of rotation, a hitherto unknown new labyrinthine reflex formed with rotatory training: the kinetic labyrinthine reflex.

Fig. 9. 0–2 seconds after the cessation of rotation to left.

Fig. 10. Eight seconds after the cessation of rotation to left.

the mid line (Fig. 4), and head-nystagmus occurred 30 times to the right for 18 seconds, i.e. the first phase of postrotatory head-nystagmus (Fig. 5). Then, the deviation shifted again to about 120° to the right and 5 jerks of head-nystagmus occurred to the left for 10 seconds, i.e. the second phase (Fig. 6). There was a certain interval between two phases. This interval was named the postrotatory head-nystagmus pause, and in this case was 7 seconds.

Results of test rotation to the right in the same animal.—The same results were obtained except that: perrotatoric head-nystagmus ceased in 9 seconds, deviation of the head was 90°–120° to the left after that, the first phase of the postrotatory head-nystagmus was 27 times to the left for 18 seconds, the second phase was 8 times to the right for 12 seconds after a pause of 4 seconds.

Results of the test in the first animal after repeated rotations for two weeks with rotation to the left.—The head and neck deviated about 60° to the right directly after the start of the test rotation (Fig. 7). Here, perrotatoric head-nystagmus continued for 5 seconds to the left and the head and neck returned to the mid line in 6 seconds. Eight seconds after the start, the head deviated about 60° in the same direction as the test rotation: this had never been observed before the repeated rotations (Fig. 8). The animal kept this deviation to the end of the rotation after that. When the rotation ended, deviation to the left increased a little, but the animal maintained his posture (Fig. 9). The first phase of postrotatory head-nystagmus took place here 9 times to the right for 4 seconds during which the head returned to the mid line. Then, the head deviated about 45° again to the right for 4 seconds (Fig. 10). There was no second phase of postrotatory head-nystagmus. That is, postrotatory head-nystagmus decreased markedly both in duration and number of jerks, as compared with that before the repeated rotations (Table 1).

Results of the test in the same animal after rotation to the right.—Initial deviation was about 45° to the left. Duration of perrotatoric head-nystagmus was 5 seconds. Deviation in the same direction occurred 7 seconds after the start at about

TABLE 1. *Changes in postrotatory head-nystagmus through training in two leghorns.*

| | | The first phase | | | The second phase | |
	Direction of rotation	Duration (min.)	jerks of nystagmus	Head-nystagmus pause	Duration (min.)	jerks of nystagmus
The first leghorn						
Before	to left	18	30	7	10	5
training	to right	18	27	4	12	8
After	to left	4	9	4[a]	0	0
training	to right	4	10	3	1	1
The second leghorn						
Before	to left	17	26	3	5	3
training	to right	18	30	7	7	4
After	to left	9	16	4	0	0
training	to right	10	17	3	0	0

[a] When the second phase of postrotatory head-nystagmus was not observed, the time from the end of the first phase to the finishing of shift to the opposite direction against the rotation was shown.

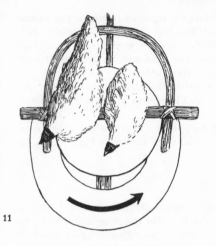

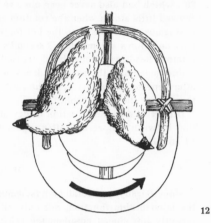

11

12

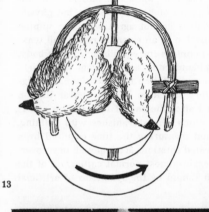

13

Trained (right) and *Untrained* (left).

FIG. 11. Two seconds after the beginning of rotation. Both heads deviate in the opposite direction of rotation: the static labyrinthine reflex (normal devitation).

FIG. 12. Eight seconds after the beginning of rotation. The head of the trained turns in the same direction of rotation through the kinetic labyrinthine reflex, while the head of the untrained deviates still in the opposite direction of rotation by the static labyrinthine reflex (normal deviation).

FIG. 13. Sixteen seconds after the beginning of rotation. The same tendency as in Fig. 12.

FIG. 14. 0–2 seconds after the cessation of rotation to left.

FIG. 15. Eight seconds after the cessation of rotation to left.

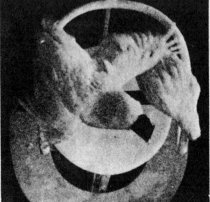

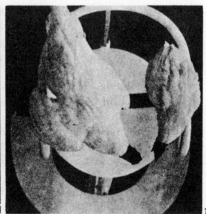

14

15

80°, which had also never been observed before the repeated rotations. The animal showed little ataxia when the rotation ended. The first phase of postrotatory head-nystagmus was 10 times to the left for 4 seconds, and the second phase took place only once for a second after 3 second's pause. The jerks and the duration of postrotatory head-nystagmus also decreased markedly (Table 1).

Fundamentally the same tendency was noted in the second animal before and after repeated rotations as in the first animal. Changes in postrotatory head-nystagmus are shown in Table 1.

Differences in the direction of deviation and posture at five stages of test rotation between, before and after repeated rotations are shown in Figs. 11, 12, 13, 14 and 15.

DISCUSSION

Since Bárány, postrotatory nystagmus has been considered a good indicator of the labyrinthine function, not only clinically but also physiologically. With the recently introduced cupulometry (13), postrotatory sensation is again called in question as a labyrinthine sensation. Indeed, a rotation test is the best method clinically for determination of labyrinthine function, for it can indicate dramatically a damaged labyrinth by showing anystagmus or hyponystagmus, but physiologically there is another question. The writers cannot agree with the opinion that regards postrotatory nystagmus and sensation as an essential reflex and sense in the labyrinth the main role of which is to maintain the equilibrium of the body. Postrotatory nystagmus is nothing but a phenomenon arising from a transient artificial labyrinthine ataxia caused by repeated rotation to which human beings and animals are not experienced. Postrotatory sensation is a vertigo or illusion by which one feels oneself still being rotated after the end of the rotation and is like a shadow of postrotatory nystagmus which almost vanishes with training. In other words, the rotation test is a method measuring the time required by a subject on whom unexperienced passive repeated rotations are imposed, in recovering from transient labyrinthine ataxia to regain almost normal eqilibrium of the body. Postrotatory nystagmus and sensation should be the symptoms of artificial Ménière's disease.

At the beginning of the test rotation, the head and neck of the animal deviated in the opposte direction to the rotation. This deviation is regarded as an important labyrinthine reflex, generally called normal deviation (11) or Kompensatorische Gegendrehung and must be a labyrinthine reflex by which the animal tries to maintain its original position before rotation. But "static labyrinthine reflex" may be a better name since the stimulus given is from standstill to rotation. As the rotation continues, stimuli received by the animal become quite dynamic. Animals are not accustomed to such passive rotation, and the deviation to the opposite direction becomes more and more marked under centrifugal force, while perrotatoric nystagmus takes place in response to the rotation, and finally the animal has difficulty in keeping itself on the perch.

After repeated rotations, the head and neck also deviated to the opposite direction to the rotation (normal deviation) at the beginning of the test rotation. However, the head turns soon against the centrifugal force over the mid line in the same direction as the test rotation after the normal deviation. This fact is adequately called the "kinetic labyrinthine reflex". Being given the same stimuli before and after the repiated rotations the heads of the trained and untrained animals deviated in different directions (Figs. 12 and 13). In other words, the labyrinthine reflex during rotation has been completely changed by repeated rotations. The writers would like to infer here that a functionally higher labyrinthine reflex has been established, for, with deviation, in the kinetic labyrinthine reflex the animal turns its head actively responding to the centrifugal force and is in full dynamic equilibrium without any ataxic movement. Moreover, the animal after repeated rotations shows scarcely any ataxic movement at the end of the test rotation with only slight change in the deviation of the head and less marked postrotatory head-nystagmus. This is why the writers named the repeated rotations "training".

Anystagmus or hyponystagmus also occur in human beings whose labyrinthine function is destroyed or decreased. But with eyes closed, they have a positive Romberg's phenomenon, cannot stand on one foot and falls down from Von Stein's goniometer with an angle of 2°–3°. An animal whose labyrinthine function is restricted shows on the one hand anystagmus or hyponystagmus, but never a kinetic labyrinthine reflex, and on the other hand normal deviation. In other words, the same hyponystagmus has a different meaning in each case. Trained persons and athletes, especially dancers, skaters and ballerinas who have superior equilibrating function, must have become free from the transient ataxia or vertigo after the Bárány's rotation test. Experiments in this paper suggest that postrotatory nystagmus and sensation produced by a normal labyrinth are due to a temporary ataxia or vertigo whose decrease indicates improvement of equilibrating function.

Güttich (12) has observed that the position of eyes, head and body of human beings is symmetrical in active and passive rotations.

The difference between activity and passivity in rotation is significant. The stimulation used in the study of labyrinthine physiology is at present passive rotation without exception. It is worthy of notice here that movements of animals especially mammals, are generally active ones. In other words, passive rotation is an unusual stimulus for mammals however mild or subliminal it may be. This is one reason why the writers suppose that temporary labyrinthine ataxia or vertigo occurs after Bárány's rotation as postrotatory nystagmus or sensation. It is presumed that man is the first mammal to have imposed passive movement on himself. Beginning on the back of a horse, vehicles have been developed to the aeroplane. The physiology of the labyrinth is intimately connected with the optic organ to favour active movement regulating muscular tone at every moment. Passive movements are new stimuli to which at first the labyrinth cannot accomodate itself, and motion sickness may be produced.

The eyes and head are deviated in the direction of rotation in the former and they left behind in the latter (normal deviation). It is surprising to find that the deviation in the direction of rotation essential to active rotation is caused by passive rotation in the trained animal through the kinetic labyrinthine reflex When an animal begins to be rotated passively, its posture shows so-called normal deviation in an effort to maintain the original position, accompanied by transient nystagmus to the direction of rotation. This reflex alone has hitherto been considered as belonging to labyrinthine reflex during rotation. However, it is apparent from the above results that a new labyrinthine reflex is formed by means of repeated rotations, which consists of deviation of the head in the direction of rotation and of its fixation in this position during rotation. This fact means that the reflex which has been considered as unchangeble changes by means of repeated stimuli to form a new reflex causing quite an inverse posture. Important, however, is the fact that even in well trained animals the head always deviates at first in the direction opposite to rotation (normal deviation) at the beginning, then deviates in the direction of rotation and fixed in this position due to the new labyrinthine reflex of higher order, i.e. the kinetic labyrinthine reflex.

Moreover, postrotatory nystagmus decreases markedly with repetition of rotation as shown in Table 1.

In short, repeated rotations can cause the formation of a new kinetic labyrinthine reflex with resultant improvement of the dynamic equilibrating function during rotation, and decrease in postrotatory nystagmus accompanied by decrease in duration of transient labyrinthine ataxia (disturbance of equilibrium).

In conclusion, decrease in postrotatory nystagmus due to repeated rotations cannot be explained by fatigue, damage or paresis but it should be understood as an improvement of the labyrinthine equilibrating function as evidenced by the establishment of the kinetic labyrinthine function.

REFERENCES

1. GRIFFITH: The decrease of after-nystagmus during repeated rotation. *Laryngoscope*, *30*, 129 (1920).
2. GRIFFITH: Concerning the effect of repeated rotation upon nystagmus. *Laryngoscope*, *30*, 22 (1920).
3. MAXWELL: The effect of habituation on the rotatory-nysta as compared with the after-nysta in the rabbit. *Amer. J. Physiol.*, *68*, 125 (1924).
4. FISCHER and BABCOCK: The reliability of the nystagmus test. *J. A.M.A.*, 72, 774 (1919).
4. DODGE: Habitutation to rotation. *J. Exp. Psychol.*, *6*, 1 (1923).
6. JONGKEES: Fortschritte der Hals-Nasen-Ohrenheilkunde. 1. Basel, S. Karger, 1953.
7. HALSTEAD, YACROZYNSKI and FEARING: Further evidence of cerebellar influence in the habitutation of after-nystagmus in pigeon. *Amer. J. Physiol.*, *120*, 350 (1957).
8. FUKUDA: Stato-Kinetic Reflexes in Equilibrium and Movement. Tokyo, Igakushoin Ltd, 1957 (in Japanese).
9. NARITA: An acquisition of aireal restistancy of the equilibrium apparatus by means of special gymnastic exercises. *Kyoto, Zibi Rinsyo, 47*, 859 (1954) (in Japanese).

10. Hoshino and Fukuda: On the postrotatory nystagmus of school boys trained with special rotatory exercises. *Kyoto, Zibi Rinsyo, 47,* 769 (1954) (in Japanese).
11. Rüedi et al.: Toxische Wirkung des Streptomycins. *Acta Oto-laryng.,* Suppl. *78,* 68 (1948).
12. Güttich: Über den Antagonismus der Hals- und Bogengangsreflexe. *Arch. Ohren-usw. Heilkunde, 147,* 1 (1940).
13. Van Egmond: Cupulometrie. *Pract. Oto-rhino-laryng., 17,* 206 (1955).

Tsukasamachi 40, Gifu

Received June 30, 1958

A NEW ARRANGEMENT OF THE VESTIBULAR EXAMINATION*

TADASHI FUKUDA and TAKASHI TOKITA
Gifu, Japan

*From the Department of Otolaryngology (Head: Prof. T. Fukuda, M.D.)
Gifu Medical School, Gifu*

The authors propose a new method of vestibular examination, testing not only the experimental nystagmus but also the righting reflexes as well as the deviation phenomenon. Experimental nystagmus is an excellent method for the determination of labyrinthine function because it can indicate dramatically a damaged or destroyed labyrinth by showing diminished nystagmus or absent nystagmus. The experimental nystagmus, however, is not a natural function of the labyrinth but a symptom of artificial labyrinthine ataxia caused by an unexpectedly intense stimulation; therefore, the authors suggest that in addition to the experimental nystagmus, the righting reflexes i. e., the natural reaction of the labyrinth necessary to maintain the equilibrium of the body and the deviation phenomenon i. e. the natural expression of the latent imbalance of both the labyrinths must be examined and we propose the adoption of the following new methods;

1. Tests related to the labyrinthine righting reflexes.
   Balance tests:
   > Von Stein's goniometer test
   > McNally's tilting table test
   > Romberg's phenomenon, standing on one foot, Mann's test etc.
2. Tests related to spontaneous or positional deviation.
   Imbalance test:
   > Nystagmus
   > The pointing test, The arm tonus reaction (Wodak, Fischer)
   > Fukuda's writing test
   > The stepping test etc.
3. Tests related to experimental nystagmus
   > Postrotatory nystagmus
   > Caloric nystagmus
   > Cupulometry

By the above new method of vestibular examination, it is possible to detect objectively subtle changes in vestibular function such as its improvement of degeneration during the course of Méniè's disease or the variation in its condition in patients before and after tympanoplasty.

We should like to propose a new arrangement of the vestibular examination not only by testing the experimental eye nystagmus but also by including in this examination the righting reflex as well as the deviation phenomenon, in accordance with the natural physiology of the vestibular labyrinth.

---

* This paper was read at the 7th International Congress of Otorhinolaryngology which was held in Paris on the 27th July, 1961

An experimental nystagmus is an excellent method for the determination of the vestibular function because by showing either hyponystagmus or anystagmus it dramatically can indicate a damaged or abolished labyrinth.   An experimental nystagmus, however, is not a natural function of the labyrinth but a symptom of artificial labyrinthine ataxia caused by an unexpected intense stimulation; therefore, we firmly believe that in addition to tests related to an experimental nystagmus, the righting reflex, i.e. the natural labyrinthine function necessary to maintain the equilibrium of the body, and the deviation phenomenon, i.e. the natural expression of the latent imbalance of both the labyrinths, must be examined.

A subject who has an impairment of his vestibular function generally complains of the sensation of dizziness and, in fact, demonstrates this sensation by his inability to stand erect.   Let us analyze this labyrinthine impairment.   In the manifestation of Romberg's phenomenon (Fig. 1) which is caused by a labyrinthine impairment the subject tends to fall toward a definite direction; however, the subject does not lose consciousness and fall to floor as would a stick.   The head and trunk which incline toward a definite direction, a splitsecond later return to the normal original position and this movement cycle is repeated again and again, thus the subject's head and trunk oscilate and he complains of dizziness.   Therefore, dizziness caused by a labyrinthine impairment can be analyzed as a deviation (the inclination toward a definite direction) and a righting motion (the return to the normal position).   This righting motion, we should like to attribute to the labyrinthine righting reflex.

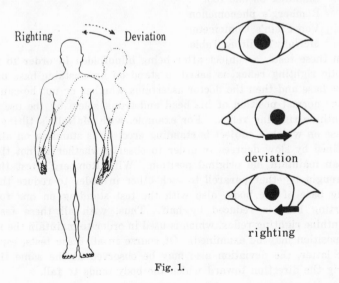

Fig. 1.

For example, in a labyrinthine eye nystagmus at first the eye-balls deviate from the normal mid-position relatively slowly but an instant later return much more quickly to the original mid-position, thus, a slow and rapid phase in this cycle may be observed to exist. These slow and rapid movements are repeated rhythmically to form a labyrinthine nystagmus and it may be stated further that the slow phase is the manifestation of the deviation and the rapid phase, the manifestation of the righting reflex.

Therefore, a labyrinthine ataxia should be examined by thinking of it as two separate functions and performing one test for the deviation and another test for that of the righting reflex. Thus, we believe that the impairment of the labyrinthine function should be and can be examined more naturally, without the addition of any excessive stimuli and we propose that the following arrangement be adopted.

1. Righting reflex
2. Deviation
3. Experimental nystagmus

Through the use of this new arrangement of the vestibular examination it is possible to observe and relate many interesting phenomena; however, since time does not permit me to lecture in great detail on each of the three parts of the new arrangement, we shall confine our lecture to tests related to the righting reflex and deviation with special emphasis on deviation.

1. *Tests related to the labyrinthine righting reflex (Balance Test)*
   Standing on one foot
   Romberg's phenomenon
   Von Stein's Goniometer
   McNally's Tilting Table

In these tests, a subject after being blindfolded in order to nullify his optic righting reflex, is asked to stand on an unstable base or on a narrow base and then the doctor ascertains whether or not he can maintain the normal position of his head and trunk solely by the use of the labyrinthine righting reflex. For example, with McNally's tilting table, the base on which a subject is standing erect or is standing on all fours is inclined by slow degrees in order to observe whether or not the subject can maintain his original position. With Romberg's test the feet are brought together pararell to each other in order to reduce the supporting base of the body, also with the test standing on one foot the supporting base is reduced by half. Thus, with all these tests the labyrinthine righting reflex, which is used in order to maintain the normal body position may be examined. Of course in all these tests, especially in the latter, the deviation also may be observed at the same time by noticing the direction toward which the body tends to fall.

2. *Tests related to spontaneous and positional deviation (Imbalance Test)*

Nystagmus
The pointing test
Arm tonus reaction
Fukuda's writing test
Fukuda's stepping test (Unterberger's Tretversuch)

A spontaneous or a positional eye nystagmus is manifest evidence of the imbalance between both labyrinths i.e. the labyrinthine ataxia. In cases where such nystagmus cannot be, or hardly can be observed, tests related to the vestibulo-spinal pathway must be adopted since with such tests the latent imbalance between both labyrinths often can be shown dramatically, especially with our blindfolded vertical writing test as well as with the stepping test, a test which we recently devised from Unterberger's Tretversuch.

Fukuda's vertical writing test

The test consists of observing the deviation in a column of letters written vertically with the eyes blindfolded and with the arm kept free from any contact with a desk, etc. The pure objectivity of this test is characterized by its results. Any deviation in the writing reveal the fact that an imbalance exists between both labyrinthine functions and this imbalance can be evoked by physiological as well as clinical stimuli to the labyrinths that are too weak to induce a nystagmus.

For example, during the interval period of Ménière's disease, when the patient wrote his name vertically, Japanese style, with his eyes open, every character showed no deviation, however, when he wrote the same characters blindfolded, every character in each column showed marked deviation every time the test was repeated, even though spontaneous nystagmus/as well as past-pointing and Romberg's phenomenon could not be observed.

Furthermore, this test easily and conveniently may be performed in bed. Therefore, the labyrinthine function of a subject even directly after a tympanoplasty may be examined visually as shown in Fig. 2, by getting the subject to write his name or any other convenient characters on a sheet of paper. We should like to emphasize here, that the patient graphically reproduces his own disease, in this case the latent imbalance between both labyrinths, on paper, not only as a clinical record for the attending physicians but also as a visual indication for the patient himself.

The stepping test

This test is a modification of Unterberger's Tretversuch and is superior to the gait test in the examination of the labyrinthine deviation phenomenon as expressed by the lower extremities.

A subject with his eyes blindfolded is asked to stand in the center of a circle which has been drawn on the floor and to stretch out both arms straight before him, then to step by repeatedly raising and lowering first one knee and then the other at a normal walking speed for a total of 50 or 100 steps while attempting to remain at his original starting position.

When there is a latent difference between both labyrinthine functions, the subject gradually turns around on his longitudinal axis in a definite direction, although he is unaware of this turning phenomenon.

When rotation or displacement of the body occur during the stepping test, this phenomenon is called stepping deviation and its parameters are expressed by the angle of rotation, angle of displacement and distance of displacement as we had shown in a paper (Fukuda, 1959).

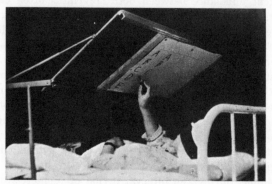

Fig. 2.

Rather recently, we have gotten a subject to perform the writing as well as the stepping tests while assuming several different positions of the head. That is, at first each test is performed with the head in the normal upright position as in the orthodox method. Then the same test is repeated first with the head inclined to the right and then to the left, both anteriorly and posteriorly in succession. Thus we have been able to show clearly the manifestation of a latent positional imbalance between both labyrinths by the upper as well as the lower extremities even when positional eye nystagmus could not be evidenced.

3. *Tests related to an experimental eye nystagmus*
        Caloric nystagmus
        Postrotatory nystagmus
        Cupulometry (The subliminal rotation)

By the above new arrangement of the vestibular examination, it is possible to detect objectively subtle changes in the vestibular function such as its improvement or degeneration during the course of Ménière's disease or the variations of its changes in a patient before and after tympanoplasty.

Case Reports:
Case No. 1

Fig. 3 shows the writing of a 22 year old male who was suffering from otitis media purulenta chronica dextra. The patient had a 40 to 50 decibel conduction loss of hearing in his right ear; however, he did not complain of any dizziness and neither spontaneous nor positional eye nystagmus could be observed. The left column of characters shows the patient's style of writing with his eyes open after he had been admitted to our clinic, whereas the other three columns show his style of writing

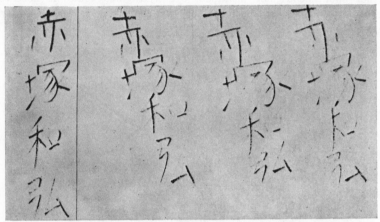

Fig. 3.

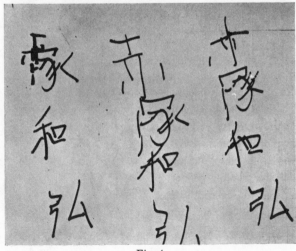

Fig. 4.

after he had been blindfolded. Please notice the marked deviation to the right of 19 to 21 degrees; however, at this time no marked derangement could be detected. This clinical manifestation showed an imbalance of both labyrinthine functions that otherwise could not have been detected.

Fig. 4 shows the style of writing of the same patient the 2nd day after a tympanoplasty had been performed. During the patient's operation it was found that his incus and malleus had been destroyed through granulation and that only his stapes remained intact; therefore, a skin graft was performed, with the graft being attached to the stapes. After the operation the patient complained of slight dizziness; however, still neither spontaneous nor positional eye nystagmus could be observed. Please notice a marked derangement of the characters as well as a marked deviation to the right.

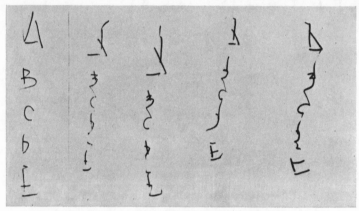

Fig. 5.

Fig. 5 shows the derangement of the letters as well as their marked deviation to the left of another patient written with eyes blindfolded the 1st day after a tympanoplasty, although the left column of letters written with eyes open at the same time shows neither deviation nor derangement. This patient also complained of dizziness but as with the preceding patient, neither spontaneous nor positional eye nystagmus could be observed.

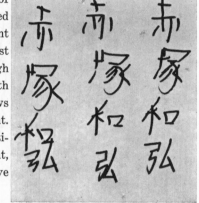

Fig. 6.

Fig. 6 shows the type of writing of the first patient, while being blindfolded the 26th day after his tympanoplasty. Please notice that the complete cure of the patient is evidenced by the fact that now his characters are written vertically and without any derangement in all three columns.

In addition, the patient's hearing was improved to where he had only a 20 to 30 decibel conduction loss of hearing in his right ear.

Case No. 2

Fig. 7 shows the writing of a 9 year old boy who was suffering from otitis media purulenta chronica sinistra. This patient had a 30 to 40 decibel conduction loss of hearing in his left ear; however, as with the first patient he did not complain of any dizziness and neither spontaneous nor positional eye nystagmus could be observed. The left column of characters shows the patient's style of writing with his eyes open, after he had been admitted to our clinic, whereas the other three columns show his style of writing after he had been blindfolded. Please again notice the deviation, but this time to the right of 10 to 12 degrees and as before no marked derangement could be detected. In most cases, the deviation is usually directed toward the ear which is impaired, however, in this case, the deviation was directed toward the unimpaired ear. This clinical manifestation showed an imbalance of both labyrinthine functions that otherwise could not have been detected.

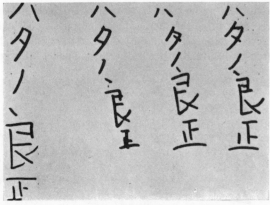

Fig. 7.

Fig. 8 shows the style of writing of the same patient the 1st day after a tympanoplasty had been performed. In this case it was found that the incus, malleus and stapes were healthy but the middle ear cavity was filled with dirty granulation and puss, and this was removed. After the operation the patient complained of slight dizziness; however, neither

spontaneous nor positional eye nystagmus could be detected. Please notice
a marked derangement of the characters as well as a marked deviation
of from 20 to 25 degrees to the right.

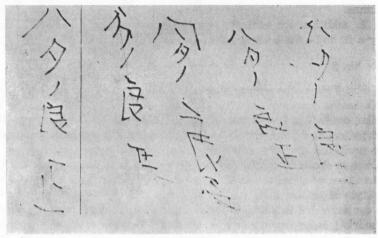

Case No. 3

Fig. 8.

Fig. 9 shows the writing of a 28 year old female during an interval
between attacks of Ménière's disease. This disease was accompanied
with tinnitus and the patient had a 20 decibel conduction loss of hearing
in her left ear. Again neither spontaneous nor positional eye nystagmus
could be observed. The left column of characters shows the patient's
name as written by her in Kanji with her eyes open. This column shows

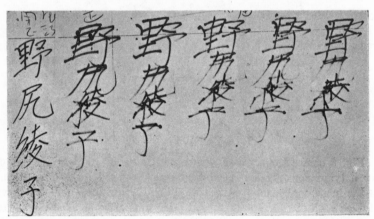

Fig. 9.

Acta oto-laryng. Suppl. 179

Fig. 10.

no deviation while the other five columns of the same characters which she wrote with her eyes blindfolded show marked deviation to the left of 18–20 degrees.   Again the latent imbalance between both labyrinthine functions clearly is evidenced by this marked deviation.

Fig. 10 shows the patient's name in five seperate columns as written by her in Kanji with her eyes blindfolded after she had recieved 20 left stellate blockades, one each day for 20 days, during the interval between attacks.   The complete disappearance of the previous deviation evidences and strongly supports the fact that a cure had been effected.

Case No. 4

Fig. 11 also shows the writing of a 25 year old female during an interval between attacks of Ménière's disease.   As in the previous case

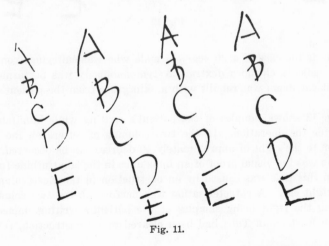

Fig. 11.

this disease was accompanied with tinnitus but this time the patient had a 20 decibel conduction loss of hearing in her right ear. As before neither spontaneous nor positional eye nystagmus could be evidenced. The four columns of letters which she wrote with her eyes blindfolded show marked deviation to the right of 20 to 25 degrees, and as before, the latent imbalance between both labyrinthine functions may be demonstrated by this deviation.

Fig. 12 shows six columns of letters. The column at the left was written by the patient with her eyes open after she had received 10 right stellate blockades one each day for 10 days, during the interval between attacks, while the other five columns were written by her after she had been blindfolded. As before, the complete disappearance of the previous deviation evidences and strongly supports the fact that a cure had been effected.

Fig. 12.

Case No. 5

This is the case of a 25 year old male who was suffering from otitis media purulenta chronica dextra. A tympanoplasty was performed and the right ear drum was rebuilt with a skin graft from the patient's right thigh.

Fig. 13 shows samples of the patient's writing while blindfolded 30 days after the operation. In the two columns of letters on the left, a deviation to the right of approximately 10 degrees can be observed. This deviation was the visual proof of an imbalance in the labyrinthine function which in this case was caused by an obstruction in the Eustachian Tube of the right ear. A catheterization was performed, the two columns of letters on the right being samples of the patient's writing immediately after the Eustachian Tube had been cleared of its obstruction, relieving

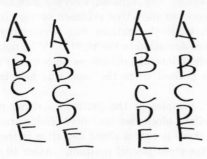

Fig. 13.

the imbalance in the labyrinthe functions. Thus by the use of the writing test it is possible to detect an imbalance in the labyrinthine functions of even an artificially created ear.

Case No. 6

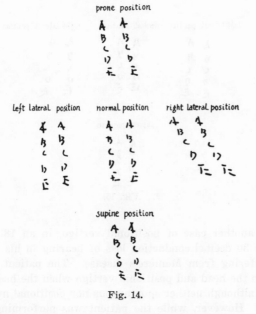

Fig. 14.

This is a case of positional vertigo in a 27 year old female with a 20 decibel conduction loss of hearing in the right ear who was suffering from Ménière's disease. When the patient's head inclined to the right slight positional vertigo was induced, whereas manifest spontaneous and positional nystagmus, an objective evidence of vertigo, could not be observed. However, while the patient was performing the writing test, an especially marked deviation to the right could be noticed when the head was held in the right lateral position as shown in Fig. 14, whereas no deviation could be noticed when the head was held in any of the other positions.

Fig. 15 shows samples of the patient's writing in all the positions after 10 right stellate blockades had been administered to the patient. Please notice that there is only a slight deviation in the writing, when the head was held in the right lateral position. After 10 more right stellate blockades the patient was completely cured as could be evidenced by the writing test for after 20 blockades no deviation in any of the head positions could be observed.

prone position

left lateral position   normal position   right lateral position

supine position

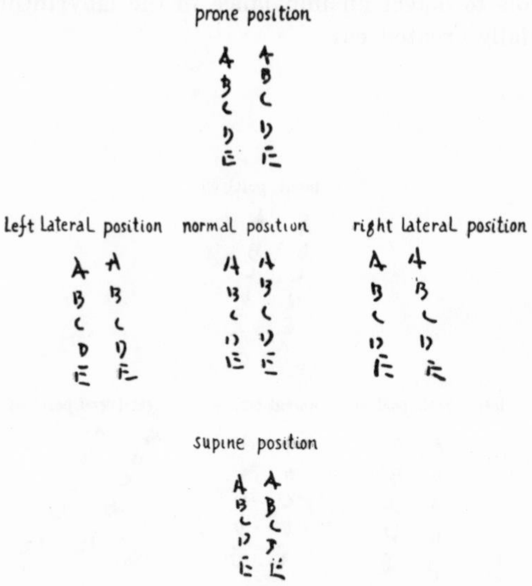

Fig. 15.

Case No. 7

This is another case of positional vertigo, in an 18 year old male with a 25 to 30 decibel conduction loss of hearing in his right ear, who also was suffering from Ménière's disease. The patient complained of discomfort in the head and positional vertigo when the head was inclined to the right, although neither spontaneous nor positional nystagmus could be observed. However, while the patient was performing the stepping test slight deviation to the right in all head positions could be observed but when the head was inclined to the right, marked deviation with ataxic steps was observed as shown in Fig. 16.

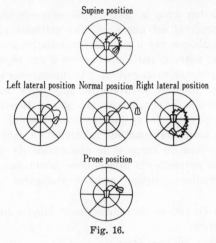

Fig. 16.

Fig. 17 shows the results of the stepping test during treatment by right stellate blockades. Please notice the marked deviation with ataxic steps when the head was inclined in the right lateral position. A complete cure was effected after 10 right stellate blockades as evidenced by further stepping tests thus this test also provides the doctor with an objective evidence of positional vertigo.

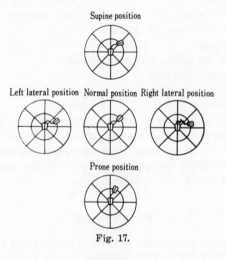

Fig. 17.

RESUMÉ

L'auteur propose une nouvelle procédure pour l'examen vestibulaire qui consisterait à examiner non seulement le nystagmus expérimental, mais aussi le réflexe de redressement ainsi qui le phénomène de la déviation, en relation avec la physiologie naturelle du labyrinthe. Le nystagmus expérimental est une excellente méthode pour une étude de la fonction labyrinthique, car il peut montrer clairement qu'un labyrinthe est endommagé ou détruit, par l'observation d'une réponse déficitaire ou par l'absence de réponse nystagmique. Le nystagmus expérimental n'est cependant pas une fonction naturelle du labyrinthe, mais un symptôme de l'ataxie labyrinthique, causée par une stimulation intense imprévue; l'auteur suggère qu'outre le nystagmus expérimental, le réflexe de redressement, c'est-à-dire la fonction naturelle du labyrinthe nécessaire à maintenir l'équilibre du corps et le phénomène de la déviation, c'est-à-dire l'expression naturelle du déséquilibre latent des deux labyrinthes, devraient être étudiés; l'auteur propose l'adoption de cette nouvelle procédure:

1. Tests relatifs au réflexe de redressement labyrinthique.
   Tests d'équilibre:
   > Goniomètre de Von Stein
   > Table de McNally
   > Phénomène de Romberg; debout sur une jambe, essais de Mann, etc.
2. Tests relatifs à la déviation;
   Tests de déséquilibre:
   > Nystagmus
   > Test de l'index; réactions de Wodak, Fischer.
   > Test de l'écriture (Fukuda).
   > Test du pas.
3. Tests relatifs au nystagmus expérimental:
   > Nystagmus calorique
   > Nystagmus post-rotatoire
   > Cupulométrie.

Par cette méthode nouvelle d'examen vestibulaire, il serait possible de découvrir objectivement des modifications discrètes de la réflectivité vestibulaire telles qu'une augmentation ou qu'une diminution au cours de la maladie de Menière; on pourrait apprécier les variations chez un malade, avant et après une tympanoplastie.

REFERENCES

WODAK, E., 1960: Einige Bemerkungen zu Fukuda's Stepping Test und Vertical Writing Test. O.R.L. Clinic, Japan (Zibi Rinsyo) 53:1.

WODAK, E., 1961: Die klinische Vestibularisuntersuchungen im Lichte neuer Methoden. Pract. Oto-rhino-laryng., 22:397.

WODAK. E., 1961: Störungen des Vestibularapparates in der täglichen Praxis. Mschr. Ohrenheilk., 95:501.

FUKUDA, T., 1959: Vertical writing with eyes covered. A new test of vestibulo-spinal reaction. Acta oto-laryng., 50:26.

FUKUDA, T., 1959: The stepping test. Two phases of the labyrinthine reflex. *Acta oto-laryng.*, *50*:95.

UNTERBERGER, S., 1938: Neue objective registrierbare Vestibularis-körper-drehreaktion, erhalten durch Treten auf der Stelle. Der "Tretversuch". *Arch. Ohr-usw. Heilk.*, *145*:478.

HIRSCH, C., 1940: A new labyrinthine reaction. The waltzing test. *Ann. of Otol.*, *49*:232.

Tsukasamachi, Gifu.

THE UNIDIRECTIONALITY OF THE LABYRINTHINE REFLEX IN RELATION TO THE UNIDIRECTIONALITY OF THE OPTOKINETIC REFLEX[1]

TADASHI FUKUDA

Gifu, Japan

From the Department of Otolaryngology, Gifu Medical School (Head: Prof. Tadashi Fukuda, M.D.)

To inquire into the physiological relationship between the unidirectionality of the labyrinthine reflex to rotatory stimulation and the unidirectionality of the optokinetic reflex five leghorns, whose eyes were covered unilaterally, were rotated with an angular acceleration of $1°/sec^2$ for a minute and the following facts were disclosed.

(1) During rotation the deficiency due to the unidirectionality of the optokinetic reflex is compensated for by unidirectional excitation of the labyrinth and perrotatory head nystagmus occurs. (2) The purpose of this compensation exists in inducing head nystagmus and thereby enabling the individual to perceive moving objects one by one and to keep the body in equilibrium during rotation. (3) In considering nystagmus from the point of view of evolution, it is inferred that the unidirectionality of the optokinetic reflex existed at first and, for the purpose of compensating for this unidirectionality, semicircular canals with their unidirectional nature evolved.

INTRODUCTION

It is well known that the rotatory stimulation of the labyrinthine organ causes perrotatory nystagmus. This reflex is provoked exclusively by excitation of that labyrinth on the side coinciding with the direction of rotation, although rotation affects both labyrinths simultaneously. This fact was first discovered by *Crum Brown* Ewald, and has since been known as the unidirectionality of the labyrinthine reflex to rotation.

· Similar unidirectionality has been reported by us in the relationship between optokinetic stimulation and optokinetic nystagmus (Fukuda & Tokita, 1957). In an animal with totally crossed optic nerves, such as a rabbit or a guinea pig, when one eye is blindfolded and when movements of the intact eye are observed in passing things one after another transversely before its face, the unidirectionality of optokinetic nystagmus is clearly seen. That is

[1] The aid of a grant from the Science Research Fund, Ministry of Education of Japan is acknowledged.

to say, the animal with one eye covered shows manifest nystagmus when the things are moved from the side of intact eye to the other side, and no nystagmus is observed when they are moved in the opposite direction.

In this report I will present and discuss the results of my investigation of the unidirectionality of the labyrinthine reflex in relation to the unidirectionality of the optokinetic reflex. I have sometimes expressed my opinion that one of the most important roles of the vestibular labyrinth is to facilitate the optokinetic reflex and to back up visual perception (Fukuda, Hinoki & Tokita, 1957). The results of the experiment to be reported here will lend further support to this opinion. The experiment revealed the fact that the unidirectionality of the labyrinthine reflex compensates for the deficiency in the unidirectionality of the optokinetic reflex. This, I believe, is a new fact found in the field of labyrinthine physiology.

In a previous paper (Fukuda & Tokita, 1957), we reported on the presence of the unidirectionality of the optokinetic reflex in animals with totally crossed optic nerves, such as rabbits, guinea pigs, etc. At that time we believed that the fact was first discovered by us. We were wrong, however, since we found later that Huizinga & Meulen (1951) had already reported on the same fact observed in a pigeon, an animal with totally crossed optic nerves. We admit here our mistake which was due to our ignorance and inattentiveness. I will present below a further step made by us based on the fact first discovered by Huizinga and Meulen.

EXPERIMENTAL

Five adult leghorns were used, for this species of animal was known to have totally crossed optic nerves and to show the unidirectionality of the optokinetic reflex clearly by manifest appearance of head nystagmus.

Experimental methods. — First, rotatory stimulation was applied according to the following procedure. The fowl was made to alight on a perch which was installed on a Bárány chair. The rotation was made with an angular acceleration of $1°/sec^2$. The rotation with an angular acceleration of $1°/sec^2$ is the so-called subliminal rotation. When the rotation with this amount of acceleration is applied to an animal of which the visual field is intercepted from the other world with a paste-board cylinder put around the animal, no perrotatory nystagmus is known to be elicited.

Regular optokinetic stimuli were produced by means of rotation of a large cylinder. The diameter of the cylinder was 2 m and the height was 1.7 m. It was hung from the ceiling. Sixteen vertical black lines each 3 cm wide were drawn at regularly placed intervals on the inner surface of the large cylinder and the fowl was located on the axis of the cylinder and this was rotated routinely for 1 minute with an angular acceleration of $1°/sec^2$, as in the subliminal rotation. (The paste-board cylinder and the large cylinder were shown in the Figs. 1 and 3 in a previous paper (Fukuda, Hinoki & Tokita, 1957).)

Experiment I.—One eye of the leghorn was covered by adhesive plaster and the visual field of the other eye was intercepted from the other world by a paste-board

TABLE 1. *Leghorns: Subliminal rotation (1°/sec²), jerks of head nystagmus per minute.*

Leghorn	Side of vision	Experiment I		Experiment II		Experiment III	
		Chair rotation to right	Chair rotation to left	Cylinder rotation to right	Cylinder rotation to left	Chair rotation to left	Chair rotation to right
No. 1	Left eye vision	0	0	50	0	54	(56)
	Right eye vision	0	0	0	53	(58)	51
No. 2	Left eye vision	0	0	66	0	55	(62)
	Right eye vision	0	0	0	55	(54)	59
No. 3	Left eye vision	0	0	41	3	40	(50)
	Right eye vision	0	0	0	40	(54)	43
No. 4	Left eye vision	0	0	53	2	54	(65)
	Right eye vision	0	0	0	63	(58)	46
No. 5	Left eye vision	0	0	51	3	52	(59)
	Right eye vision	0	0	0	57	(55)	53

cylinder put around the fowl. Then rotation of the Bárány chair with an angular acceleration of 1°/sec² was imposed on the fowl for a minute. The results obtained in five leghorns were exactly the same as the subliminal rotation theory indicated, i.e. there occurred no perrotatory head nystagmus during rotation either to the right or to the left. A leghorn with an eye covered usually showed slight tilting and turning of the head toward the side of the open eye (see item "Experiment I" in Table 1).

Experiment II.—The paste-board cylinder used in Exp. I was removed and a large cylinder was rotated for a minute with an acceleration of 1°/sec² around the fowl with one eye blindfolded. The optokinetic head nystagmus thus produced by pure optokinetic stimuli was measured. The results in a representative case (No. 1) were as follows. At first the right eye was covered and the cylinder was rotated from the left, the intact eye side, to the right, the covered eye side. When the cylinder began to rotate the fowl's head gradually deviated to the right, followed by rapid return to the original head position. Thus continuous head nystagmus was induced and its jerks became more and more frequent in parallel with increase in velocity of rotation, until at last the number of jerks reached 50 per minute. Next the cylinder was rotated from the right, covered eye side, to the left, the intact eye side. On this occasion the fowl did not show optokinetic nystagmus at all. The ratio of the number of jerks per minute in case of rotation of the cylinder to the left, to that in case of rotation to the right was thus 0 to 50. During these procedures the fowl was left alighting on a perch without any restraint of the head or body. During rotation of the cylinder to the left the fowl by no means stood still in the upright posture, but its head tilted and turned slightly to the side of the intact eye and, although no head nystagmus was induced by optokinetic stimuli, it sometimes showed nictating movements with an air suggestive of some embarrassment and, more rarely, moved its head in a manner suggestive of shaking something troublesome off the head. When left eye of the same leghorn was covered, rotation of the cylinder from the left, the side of the covered eye, to the right, the side of intact eye, produced no optokinetic head nystagmus, whereas rotation in the opposite direction caused brisk optokinetic head nystagmus, with as many as 53 jerks per minute.

Similar facts were observed in the four other leghorns examined. A summary of the results of the experiments on these five leghorns is shown under the item "Experiment II" in Table 1. In leghorns Nos. 3, 4 and 5 some jerks of head nystagmus—three jerks per minute in No. 3, two jerks in No. 4, and three jerks in No. 5—were observed, when the cylinder was rotated from the covered eye side to the intact eye side. These jerks, however, were some atypical ones and should rather be called nystagmoid movements. Moreover, their rapid components were often directed toward the opposite side to those in typical nystagmus, i.e. they were inversion-type nystagmus. The above experiment thus evidenced in leghorns the presence of the unidirectionality of the optokinetic reflex, which had been proved in pigeons by Huizinga and Meulen by means of head nystagmus, as well as in rabbits and guinea pigs by us by means of eye nystagmus. All of these animals have totally crossed optic nerves.

Experiment III. In contrast to the preceding experiment the large cylinder was made immobile and the chair, on which a fowl with one eye covered perched, was rotated with an angular acceleration of $1°/sec^2$ for a minute. Thus numbers of jerks of head nystagmus were counted during both right and left rotations and these were compared with each other. During the above procedure the fowl received two kinds of stimulation: one which was applied to the labyrinth by means of rotation *per se* and the other which was applied to the open eye by means of relative rotation —in the opposite direction to chair rotation—of the cylinder (which was really immobile) (Fig. 1). The former was the same stimulation as in Experiment I (the subliminal rotation) and the latter the same as in Experiment II, since the acceleration of chair rotation was $1°/sec^2$. As described in the section of Experiment I the fowl whose visual field was intercepted by a paste-board cylinder did not show perrotatory nystagmus during rotation with an acceleration of $1°/sec^2$. Based upon this and similar facts the theory of subliminal rotation and cupulometry has insisted on the physiological inactivity of the labyrinth during the subliminal rotation. If this theory were right, the rotation of the chair in Experiment III would cause no labyrinthine excitation, and optokinetic head nystagmus would be induced alone, due to relative rotation of the cylinder around the fowl. If this assumption were right, head nystagmus in Experiment III would be the same as in Experiment II and would be manifest only when the relative movement of the cylinder is from the side of the intact eye to that of the covered eye and absent when the direction of the relative movement is opposite; i.e., the unidirectionality of the optokinetic reflex would be expressed in Experiment III, too. In other words, the fowl with one eye covered should show head nystagmus only when the rotation of the chair is one of either direction, right or left, and show none when the rotation is in the opposite direction. The fact, however, is that the fowl with one eye covered demonstrated brisk head nystagmus irrespective of whether the rotation of the chair was to the right or to the left. There was no significant quantitative and qualitative difference in its briskness between rotation to the right or to the left.

A representative case (No. 1) will now be demonstrated in detail. When the right eye of the fowl was covered and the chair was rotated to the left with an acceleration of $1°/sec^2$ for a minute, relative movements of the immobile cylinder to the rotating chair were seen to act as optimal optokinetic stimuli which were to produce optokinetic nystagmus, just as in case of rotation of the cylinder around the immobile chair from the open eye side, the left, to the covered eye side, the right. The number of jerks of head nystagmus was 54 per minute. This value is grossly equal to that

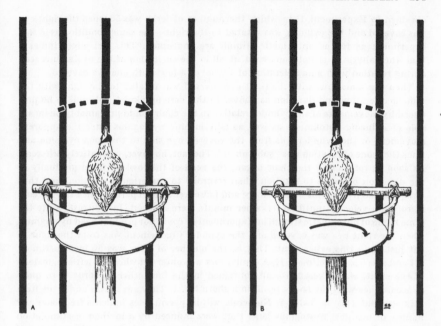

Fɪɢ. 1. Experiment III. A leghorn is placed on a perch installed on a rotating chair. Its right eye is covered with adhesive plaster and the chair is rotated with an angular acceleration of 1°/sec² (subliminal rotation). When the rotation is in the direction indicated by a solid arrow in the figure, vertical black lines drawn on the inner surface of a large cylinder and which are one meter distant from the fowl's head move relatively in the direction indicated by a broken arrow in the figure, viewed from the fowl's position. This relative movement stimulates the fowl optokinetically.

A. Left rotation of the chair. According to the subliminal rotation theory and as shown practically in Experiment I head nystagmus of labyrinthine origin cannot be induced by that mode of chair rotation. As for head nystagmus of optokinetic origin, it is supposed to be induced by the relative movement of the vertical black lines, since the movement is from the left, the open eye side, to the right, the covered eye side—the direction optimal for inducing the optokinetic reflex. The facts confirmed this assumption. As soon as the rotation began, lively head nystagmus appeared and the number of jerks per minute was the same as in Experiment II.

B. Right rotation of the chair. On this occasion relative movement of the black lines is from the right, the covered eye side, to the left, the open eye side. As shown in Experiment II this mode of optokinetic stimulation should be incapable of eliciting head nystagmus. The head nystagmus of labyrinthine origin should not be induced either, since the rotation is "subliminal". Therefore, no head nystagmus should be elicited by this mode of rotation, as inferred from the results of Experiments I and II. The facts, however, were far from this inference. As soon as rotation began, the head of the fowl deviated strongly in the direction opposite to the rotation, and lively head nystagmus ensued. This nystagmus was quite the same qualitatively and quantitatively as that seen during left rotation of the chair except the opposite direction of the rapid phase. Numerals put in parentheses in Table 1 are numbers of jerks per minute of this lively head nystagmus, which should be zero if compensatory action of the labyrinth for the unidirectionality of the optokinetic reflex were absent.

obtained in Experiment II in which the number of jerks was 50 when the right eye was covered and the cylinder was rotated to the right—the same condition as in this experiment as far as optokinetic stimuli are concerned. This fact might indicate that the labyrinth is not concerned at all in the elicitation of the nystagmus seen during rotation with an acceleration of $1°/sec^2$ of a fowl with one eye covered.

Then the same fowl with the right eye covered was rotated to the right with the same acceleration and duration as above. In this case optokinetic stimuli to be produced by movements of the cylinder relative to the chair would presumably be incapable of inducing optokinetic as well as labyrinthine nystagmus since the apparent movement of the cylinder was from the covered eye side to the open eye side and since the mode of rotation was "subliminal". The fact, however, was entirely different. As soon as rotation of the chair began, the head of the fowl deviated gradually in the direction opposite to rotation, then returned to the original position, and thus jerks of head nystagmus became more and more frequent in parallel with the increase in rotation velocity, until 56 jerks per minute were reached. In contrast, quite the same mode of optokinetic stimuli in Experiment II could not produce head nystagmus, when the right eye was covered and the cylinder's movement was from the right to left just as in this experiment. That is, the number of jerks was 56 in Experiment III and was 0 in Experiment II. A similar fact was observed when the left eye, instead of the right, was covered. Results obtained in the four other leghorns were quite in accordance with the above result in leghorn No. 1. They are shown under the item "Experiment III" in Table 1. Numerals within parentheses indicate frequency per minute of manifest nystagmus jerks that were induced by a method of stimulation which could never have induced nystagmus, according to the theory of "subliminal rotation" or to the law of unidirectionality of the optokinetic reflex.

These results lead to the following question: Why is such pronounced nystagmus induced by a combination of the two conditions, rotation and optokinetic stimuli, despite the fact that neither of these alone can induce nystagmus? In the next section I will discuss this problem and the physiological relationship existing between the unidirectionality of optokinetic nystagmus and that of the labyrinthine nystagmus.

DISCUSSION

In the first place the reason I devised this experiment will be stated. The results of experiment with five leghorns, which were described above under the heading of Experiment III, indeed evidenced the correctness of the hypothesis which I had previously advanced. Thus the hypothesis preceded the findings in this case. Let me first describe the process of thought which led to the hypothesis.

As described above, the unidirectionality of the labyrinthine reflex to rotatory stimulation has been known. Therefore, when I found the unidirectionality law in the optokinetic reflex in animals with totally crossed optic nerves, such as rabbits, guinea pigs, etc., I simply thought that this unidirectionality was analogous to that of the labyrinthine reflex and that the former was subordinate to the latter. After careful consideration, however, I came to the conclusion that the unidirectionality of the optokinetic reflex existed primarily and the vestibular reflex with a unilateral nature has evolved secondarily to compensate for deficiencies due to the unidirectionality of the optokinetic reflex; i.e., as concerns the unidirectionality law the labyrinthine reflex is subordinate to the optokinetic reflex.

One may well understand the correctness of this opinion if one considers the physiology of nystagmus. It is well known that vestibular stimulation can induce eye nystagmus, and this has been used as one of the best indicators in testing labyrinthine function. Investigation of vestibular function has hitherto centered on this phenomenon specifically.

I dare say, however, that one important problem has generally been overlooked in these investigations: that is, the physiological mechanism of the phenomenon that the excitation of the labyrinth, the organ for bodily equilibrium, does activate the eye balls, the organs for visual sense. Taking no heed of this problem, but using eye nystagmus as the chief indicator of the reflex, the main efforts have hitherto been made to investigate the mechanism within the labyrinth purely, as exemplified by the device of a rotating chamber in cupulometry. Such investigations are, of course, excellent. But I believe we must direct our effort also to investigation of the basic problem hitherto missed: i.e., why the optic organs react to stimulation applied to the vestibular labyrinth. A new concept of labyrinthine physiology must be built around this problem.

Now nystagmus will be considered from the evolutionary point of view. In animals of the lower classes, such as crabs, optokinetic nystagmus can be induced by objects moving one after another before the eyes. This nystagmus is a typical one with a slow and a rapid phase. But a crab does not show any eye nystagmus no matter how rapidly it is rotated when its visual field is kept intercepted from the environment. (When the visual field is not intercepted, optokinetic nystagmus occurs because of relative movements of the environment.) This failure in appearance of perrotatory nystagmus of vestibular origin in a crab is natural in view of the fact that its vestibular labyrinth consists of otocysts alone and has no semicircular canals. This fact is very important inasmuch as it clearly shows that optokinetic nystagmus developed first and it is not until the development of the semicircular canals that the nystagmus of vestibular origin takes part in the nystagmus phenomenon. This is the reason why I consider optokinetic nystagmus a primary phenomenon and labyrinthine nystagmus a secondary one. The unidirectionality of the optokinetic reflex is, therefore, not to be explained simply by analogy to the unidirectionality of the labyrinthine reflex, but the former unidirectionality is to be explained first and as the next step the relation or contribution of the latter unidirectionality to the former is to be clarified.

The experiment reported here has been carried out under these considerations. In Experiment I "subliminal rotation" with interception of the visual field was shown to induce no perrotatory nystagmus. In Experiment II optokinetic stimulation by means of objects moving from the covered eye side to the open eye side with the same acceleration as in subliminal rotation was found to induce no optokinetic nystagmus. From these results it may be well supposed that nystagmus cannot be induced by a combination of these two kinds of stimuli, i.e. subliminal rotation of the animal with one eye covered from the open eye side to the covered eye side. However, according to my opinion that the role of the vestibular labyrinth consists in compensating for the deficiencies in the optokinetic reflex, it may rather be assumed that even the so-called subliminal rotation of the animal caused nystagmus due to compensatory function of the vestibular labyrinth for the unidirectionality of the optokinetic reflex. Experiment III has clearly shown the accuracy of this assumption. In a fowl with one eye covered subliminal rotation, in a direction that caused relative movements of the environment from the covered eye side to the open eye side, induced head nystagmus with as frequent jerks as seen during rotation in the opposite direction. I think this marked nystagmus was produced by labyrinthine function. The reason has been fully stated in a previous investigation of a rabbit with its both eighth nerves cut (Fukuda, Hinoki & Tokita, 1957). The failure to produce perrotatory nystagmus in Experiment I was due to the specific experimental condition that the visual field was intercepted from the other world. When optokinetic stimuli are applied simultaneously with rotatory stimuli the labyrinthine reflex is encouraged by them and brought into play to show manifest perrotatory nystagmus as in Experiment III.

Thus the results of the experiment here reported have proved once again that the theory of subliminal rotation is doubtful. They also revealed the true significance of the labyrinthine reflex for visual perception. That is to say, the labyrinthine reflex compensates for the deficiency in the optokinetic reflex that is seen during rotation—an abnormal condition—and gives rise to nystagmus, in reacting to the rotation in a certain direction which cannot induce nystagmus optokinetically, and in compensating for the unilateral nature of the optokinetic reflex, for the final purpose of promoting visual perception during rotation.

The significance of the appearance of nystagmus of vestibular origin as a means for compensating for the unidirectionality of the optokinetic reflex is further explained as follows. When optokinetic nystagmus is observed in man, the slower the movements of the objects before the eyes the longer are the intervals between jerks, and the faster the movements of the objects, the shorter are the intervals. But when the velocity of the moving objects exceeds a critical value, nystagmus stops and the eye balls rest immobile in the middle position. These objective findings correspond to the subjective sensations. As far as nystagmus can be observed the individual is perceiving all the moving objects one by one. When the critical velocity is exceeded and movements of the eye balls stop he becomes unable to perceive objects separately, but he sees a vague flow before his eyes. It may not be a miss to apply this relationship confirmed in human subjects to experiments in fowls.

The fact that one eye of an animal with totally crossed optic nerves can react with nystagmus to things moving before it in one direction, is to be interpreted as evidence of the animal's perception of moving things. On the other hand, the fact that the same eye cannot react to things moving in the opposite direction seems to denote that the animal is not perceiving moving things individually. When an animal is rotated, its environment rotates relatively around it in the opposite direction. It must keep its bodily equilibrium and perceive its environment with the help of the equilibrating reflex expressed in the form of nystagmus. In a leghorn with one eye covered nystagmus cannot be elicited optokinetically when the movements of objects are from the covered eye side to the intact eye side. Therefore, to obtain exact perception of the moving environment and to maintain its bodily equilibrium during rotation which causes such a mode of environmental movements as above, its labyrinth functions and provokes nystagmus. In other words, the labyrinth on that side coinciding with the direction of rotation is activated by optokinetic stimuli received by the contralateral eye and provokes nystagmus in that mode which would be seen were the optokinetic reflex not unidirectional, thereby compensating for the lost function of the ipsilateral covered eye to induce optokinetic nystagmus and provide the individual with correct visual perception and bodily equilibrium. Since the appearance of nystagmus is proved in human subjects to indicate the ability of perceiving individual moving objects, I believe in the correctness of these considerations. Thus the presence of a close physiological relationship between the unidirectionality of the optokinetic reflex and that of the labyrinthine reflex has been first disclosed here.

It is true that animals perceive environmental objects with both eyes, not with one eye alone, and thereby react to movements of objects. Animals with totally crossed optic nerves are not the exception when both eyes are open. Optokinetic stimuli or rotation in either direction, right or left, can produce manifest nystagmus with the same number of jerks in each occasion. The close relationship between two unidirectionalities as described above has been discovered in an animal with one eye covered, but it can be sure that the same relationship—compensation of deficiency in the unidirectionality of the optokinetic reflex by unidirectional action of the labyrinth—continues to exist in the state of binocular vision—the natural state. This discovery, I believe, is an important milestone in the physiology of nystagmus.

Previously we reported on the unidirectionality of the optokinetic reflex seen in other animals with totally crossed optic nerves, such as rabbits, guinea pigs, etc. In these animals we could prove the existence of similar relationship to that reported here with respect to eye nystagmus. In experimenting with these animals, however, it could be proved only qualitatively, and quantitative study failed to obtain such uniform results as seen in nystagmus jerks in leghorns. The cause of this failure, I think, is binding up the animal's head and body during experiment in order to observe eye nystagmus in its pure form. Even leghorns failed to show clear-cut results when they were bound up. Such results were obtained only when they were allowed to alight on a perch without any restraint. In other words, in a restrained posture exteroceptive and proprioceptive impulses due to the restraint will interfere with entirely normal functioning of the reflex, resulting in failure in quantitative study. This assumption is strongly supported by the fact found by us (Ichikawa, 1958; Kurata, 1958) that optokinetic as well as perrotatory head nystagmus in a leghorn is influenced greatly by its bodily posture.

ACKNOWLEDGEMENT

The author wishes to express his cordial thanks to Drs. M. Hinoki, T. Tokita, T. Kurata, S. Aoki, E. Yonekura, T. Shiraki, T. Takeuchi, Y. Ando and H. Ichikawa for the experimental assistance with which the author's hypothesis was clearly evidenced. Thanks are due to Dr. K. Sakata for assistance with the English composition.

REFERENCES

FUKUDA, T., HINOKI, M., and TOKITA, T., 1957: Provocation of labyrinthine reflex by visual stimuli. *Acta oto-laryng.*, *48*, 425.

FUKUDA, T., and TOKITA, T., 1957: Über die Beziehung der Richtung der optischen Reize usw. *Acta oto-laryng.*, *48*, 415.

HUIZINGA, E., and MEULEN, P., 1951: Vestibular rotatory and optokinetic reactions in the pigeon. *Ann. Otol. Rhino. & Laryng.*, *60*, 927.

ICHIKAWA, H., 1958: The influence of proprioceptive stimuli upon per- and postrotatory nystagmus. *Jap. J. Otol. Tokyo*, *61*, 1573 (in Japanese).

KURATA. T., 1958: Posture and optic reflex. *Jap. J. Otol. Tokyo*, *61*, 1578 (in Japanese).

Tsukasamachi 40, Gifu

Received February 28, 1958

PROVOCATION OF LABYRINTHINE REFLEX BY VISUAL STIMULI

Evaluation of the Theory of Subliminal Rotation

BY TADASHI FUKUDA, MANABI HINOKI AND TAKASHI TOKITA

GIFU, JAPAN

Introduction

R ecently, a new method of examination on the labyrinthine function called the subliminal rotation (Egmond, Arslan) has been universally put into practice in the place of Bárány's method. This method is based upon the theory which establishes a threshold of angular acceleration about the labyrinthine function when a man or an animal is rotated. The cupulometry is also founded on this theory. It insists that the labyrinth remains unstimulated if the subject is rotated at such a low angular acceleration below a certain limit as to cause no perrotatoric nystagmus. The purpose of this paper is to criticize this theory, producing counter-evidence especially in regard to the optic condition of the subliminal rotation.

Experimental

Six rabbits were used. The angular acceleration of $1°/sec^2$ is regarded subliminal when using rabbits, for the perrotatoric nystagmus is never induced in this case. The velocity was therefore accelerated $1°$ every second, and in 3 minutes it became one rotation every two seconds or the same angular velocity as Bárány's. Nystagmus was observed and recorded in ENG (electronystagmograph) for these 3 minutes. The results which follow were obtained with the rotations of $1°/sec^2$; however, the same conclusions were drawn from the series of experiments using the angular acceleration of $0.3°/sec^2$ (Egmond, de Wit).

The first rotation

The subliminal rotation, quite the same as the original one was given to rabbits, intercepting their visual fields by a smaller cylinder made of paste-board (Fig. 1 a, b).

The eyes had been kept in the middle position and no perrotatoric nystagmus was observed during the rotation for 3 minutes. From these results it seemed that the labyrinth remained unstimulated as the theory insisted, and nystagmus was caused for the first time after the rotation, also in accordance with the theory (Fig. 2). It will be shown, however, that the labyrinth is stimulated even during the rotation.

The second rotation

Regular optokinetic stimuli were added to the subliminal rotation contrary to the original method in which a man was robbed of his visual field by being made to wear a glass of 20 D (Arslan), or being seated in a rotating chamber (de Wit)

From the Department of Otolaryngology, Gifu Medical School (Head: Prof. Tadashi Fukuda, M.D.)

Fig. 1 *b*. The first rotation. The cylinder is peeped into. The same condition as cupulometry. Perrotatoric nystagmus never occurs.

Fig. 1 *a*. The first rotation.

at the cupulometry in order that the labyrinthine reflex might be observed purely. After taking away the smaller cylinder at the first rotation, the chair with a rabbit on it was put in the centre of a larger cylinder, and rotated subliminally. The larger cylinder was made of white cloth 2 m in diameter and 1.7 m in height, and hung perpendicularly from the ceiling. Along the inner surface of this cylinder 16 black lines were drawn 3 cm in width at the same intervals (Fig. 3 *a*, *b*; Figs. 4 and 6). An animal rotated on a chair at the centre of it, is bound to receive regular visual stimuli. In this case, not only the postrotatoric nystagmus but also the perrotatoric nystagmus was caused, and that continuing quite lively for 3 minutes (Fig. 5). Generally, when a man or an animal is rotated without covering up their visual field, nystagmus is undoubtedly induced. This nystagmus is caused by the labyrinthine stimulation on the one hand, and by the optokinetic stimulation on the other. If the second rotation was below the threshold and the labyrinth was left unstimulated in accordance with the theory, the labyrinthine reflex could not be induced at this rotation, so that the nystagmus observed during the 3 minutes must be caused wholly by visual reflex. Thus, the same optokinetic stimuli was designed for the next (3rd) rotation.

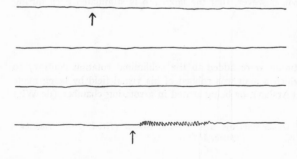

Fig. 2. ENG. The first rotation with a normal rabbit. An arrow above indicates the beginning of rotation and that below indicates cessation. Nystagmus does not occur during the rotation, but occurs after the rotation—a result in accordance with the theory of cupulometry and the subliminal rotation.

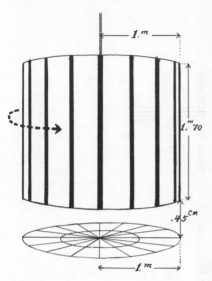

Fig. 3 a.

Fig. 3 b.

The third rotation

In this rotation the chair was fixed, and the larger cylinder around it was rotated subliminally ($1°/sec^2$) in another direction in order to give the rabbit the same optokinetic stimuli as at the second rotation (Fig. 6). Indeed, in this way the nystagmus was induced, but was much less frequent (1/60–1/40 times) than in the case of the second rotation (see Table 1, Figs. 7 and 8), and that only when the velocity was low at the earlier stage. Over the angular velocity of one rotation per ten seconds the nystagmus was not caused. Through quite the same optokinetic stimuli, the nystagmus, remarkably livelier quantitatively and qualitatively, was caused at the second rotation at which the rabbit was rotated subliminally. In fact, the nystagmus induced at the second rotation was never purely optokinetic, and

Table 1. *Rabbits: Subliminal rotation ($1°/sec^2$), jerks of nystagmus for 3 minutes.*

First rotation		Second rotation		Third rotation	
to right	to left	to right	to left	to right	to left
0	0	380	330	40	46
0	0	139	148	10	8
0	0	362	413	41	10
0	0	245	218	15	20
0	0	180	313	20	10
0	0	238	138	78	44

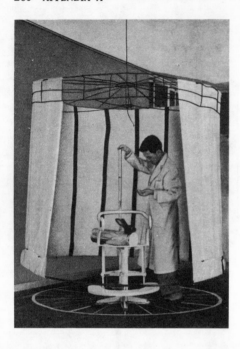

Fig. 4. The second rotation. The inner aspect of the large cylinder (partly opened). The chair is subliminally rotated.

the labyrinth must have played an important part in stirring up the nystagmus. In view of this, the labyrinth cannot be regarded as remaining unstimulated during the so-called subliminal rotation.

Substantial proof for the above-mentioned view was obtained by giving the second and third rotations to a rabbit whose labyrinthine function had been confined by cutting the vestibular nerves bilaterally. This time, the lively nystagmus previously observed at the second rotation was no longer induced. With the rabbit treated, the nystagmus at the second rotation was, quantitatively and qualitatively, nearly equal to that gained at the third rotation (Figs. 9 and 10). In other words, the nystagmus with nearly the same jerks and nature was caused either by rotating the chair or by rotating the larger cylinder, and was nothing but the pure optoki-

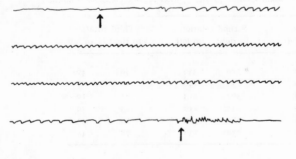

Fig. 5. ENG. The second rotation with a normal rabbit. By adding optokinetic impulse, but with entirely the same rotation of the chair as the first rotation, nystagmus takes place quite actively through the rotation. The after-nystagmus also occurs as in Fig. 2.

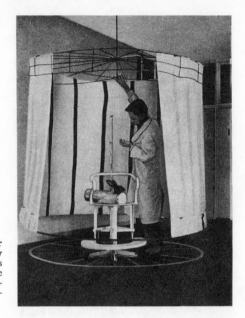

Fig. 6. The third rotation. The inner aspect of the large cylinder (partly opened). A pure optokinetic impulse has been imposed on the rabbit, rotating the large cylinder inversely to the second rotation according to the subliminal rotation.

netic nystagmus. From these results with the treated rabbit, the difference between the nystagmus which had been observed at the second and the third rotations with the untreated rabbits was regarded definitely as being caused by the labyrinthine function. (Note that the solid and the dotted lines in Fig. 11 run quite contiguously, although they are remarkably separated from each other in Fig. 8.)

Discussion

Generally, when things pass through the visual field one after another, nystagmus becomes livelier with the increase in quantity of things perceived individually, as well as in proportion to the velocity of passing things. But, when the velocity becomes higher beyond a certain limit, nystagmus ceases to be observed, so the eyes come to be fixed in the middle position: when visual stimuli are given too fast one after another, the optic organ does not react to the stimuli, and this results in a

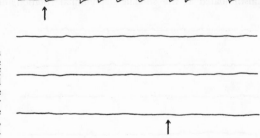

Fig. 7. ENG. The third rotation with a normal rabbit. A pure optokinetic impulse by rotation of the large cylinder. If the second rotation is subliminal and the labyrinth is neither stimulated nor reflected, the ENG's must be the same in Fig. 5 and Fig. 7. But, in fact they differ considerably.

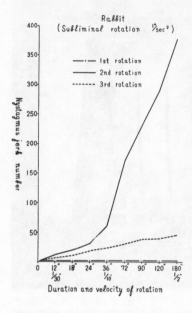

Fig. 8.

lack of the optokinetic nystagmus. In the experiment on the human being, an individual movement of things ceases to be recognized beyond such a limit, and the movement of objects before the eyes is perceived on the whole as a vague flow.

Now, the more the subject accepts the individual movement of things, the more numerous are the jerks of nystagmus, so that the nystagmus induced by an optic stimulation of higher velocity at the second rotation is nothing but a proof that the subject has recognized movements of the black lines individually. Through these facts about the optokinetic nystagmus, the quantitative and qualitative promotion of nystagmus at the second rotation has come to be thought of as being caused by the labyrinthine function which makes the rabbit catch or recognize things moving before the eyes separately as much as possible, reacting to an abnormal condition of such a passive rotation at the higher angular velocity to which the optic organ is unable to react. Thus it is clear that the labyrinth, far from remaining unstimulated at the so-called subliminal rotation, helps the optic recognition of

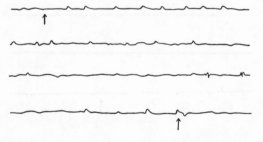

Fig. 9. ENG. The second rotation in a rabbit with its 8th nerves cut.

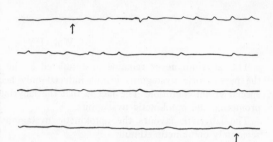

Fig. 10. ENG. The third rotation in a rabbit with its 8th nerves cut. By cutting the 8th nerves, the considerable difference between Fig. 5 and Fig. 7 disappears, and the ENG in both the second and third rotations becomes the same quantitatively and qualitatively.

things, promotes the optokinetic nystagmus, and plays quite an important part in the physiology of equilibrium.

The optic nystagmus is proved in all the animals having eyes, even in a low-developed one like a crab. But the labyrinthine nystagmus is never caused in animals having only otocysts, although a lively optokinetic nystagmus is induced. While this seems natural because they have no semicircular canals, it does show the evolution of the physiology of nystagmus, viz. that nystagmus is originally caused by optic stimuli. Semicircular canals appear as an animal develops, and the labyrinth begins to take part in stirring up nystagmus, so that the labyrinthine nystagmus must find its physiological meaning in favouring the optokinetic nystagmus. In relation to nystagmus the optic organ plays the principal role, while that of the labyrinth is secondary. It has been found from the experiments reported in this paper that the labyrinth promotes the optokinetic nystagmus even at the so-called subliminal rotation.

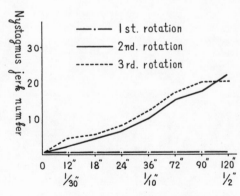

Fig. 11.   Duration and velocity of rotation

28 – 573351 *Acta Oto 48: 5–6*

SUMMARY

The labyrinth never remains unstimulated at the so-called subliminal rotation; the perrotatoric nystagmus is not induced only because the subject is under a special visual condition. The fact is that the labyrinth is stimulated and reacts in promoting the optokinetic nystagmus.

The labyrinth favours the optokinetic nystagmus by making a subject catch separately optokinetic stimuli.

Nystagmus rises originally from the optokinetic stimuli, and the labyrinth plays its physiological part secondarily, thereby favouring the optokinetic nystagmus.

REFERENCES

Arslan: On the renewing of the methodology for the stimulation of the vestibular apparatus. *Acta oto-laryng.*, Suppl. *122*, 1955.

van Egmond: Standardisation et simplification de l'examen de l'appareil vestibulaire. *Acta oto-laryng.*, *43*, 283, 1953.

— Cupulometrie. *Pract. oto-rhino-laryng.*, *17*, 206, 1955.

de Wit: Seasickness: *Acta oto-laryng.*, Suppl. *108*, 1953.

40 Tukasamachi Gifu

Received August 19, 1957

Confin. neurol. *24:* 127–139 (1964)

From the Department of Otolaryngology, Gifu Medical School
(Head: Prof. *Tadashi Fukuda*), Gifu, Japan

Diagonal Nystagmus

Influence of Optokinetic Stimulation on Postrotatory and Caloric Nystagmus

By T. FUKUDA and T. TOKITA

The functional relationship between the vestibular labyrinth and optical organ is studied on the basis of the authors' belief that one of the most important functions of the vestibular labyrinth is to facilitate the optokinetic reflex and the visual perception; this is clearly evidenced by the following.

In lower animals, such as crabs, optokinetic nystagmus can be induced by objects moving one after another before the eyes. This nystagmus is a typical one with a slow and a rapid phase. But a crab does not show any eye nystagmus no matter how rapidly it is rotated when its visual field is kept isolated from the environment (when the visual field is not isolated, optokinetic nystagmus occurs because of the relative movement of the environment). This failure in the appearance of perrotatory nystagmus of vestibular origin in a crab is natural in view of the fact that its vestibular labyrinth consists of otolith crystals alone and has no semicircular canals. This phenomenon seems to be simple but from the phylogenetic point of view, it is very important and interesting. It could be assumed that optokinetic nystagmus is a primary phenomenon and labyrinthine nystagmus is an advanced reaction after semicircular canals develop. The above observation suggests that labyrinthine nystagmus is a modifier of primarily existing optokinetic nystagmus. Under such an assumption, the following experiments were carried out.

If so-called subliminal rotation is performed so that the vestibular labyrinth is not stimulated, reflexes are apparently elicited from the vestibular labyrinth when optical stimulation is applied simultaneously (*Fukuda et al.* 1957).

It is well known that the rotatory stimulation cause perrotatory nystagmus which is elicited by excitation of the labyrinth on the side coinciding with the direction of rotation (unidirectionality of the labyrinthine reflex to rotation). Similar unidirectionality between optokinetic stimulation and optokinetic nystagmus has been observed by the authors in animals with totally crossed optic nerves (rabbits or guinea pigs), in which one eye is blindfoled. The unidirectionality of optokinetic nystagmus is clearly demonstrated when movements of the non covered eye are observed while objects are passed one after another transversely before its face. Such an animal shows definite nystagmus when the objects are moved from the side of the non-covered eye to that of the covered eye; no nystagmus appears when they are moved in the opposite direction (*Fukuda, Tokita*, 1957). One of the authors reported that the unidirectionality of the labyrinthine reflex may compensate the deficiency in the unidirectionality of the optokinetic nystagmus described above; the vestibular labyrinth facilitates the optokinetic nystagmus and increases the visual perception (*Fukuda*, 1959). Apparently the vestibular labyrinth compensates and modifies the optokinetic reflex in some way.

The above experiments deal primarily with perrotatory nystagmus. Here we would like to discuss the relationship between post-rotatory, caloric and optokinetic nystagmus. Perrotatory nystagmus facilitates and modifies optokinetic nystagmus. Postrotatory and caloric nystagmus, on the other hand, are suppressed and covered by optokinetic nystagmus and follow its pattern.

As far as the optic-vestibular coordination is concerned many studies have been carried out *(Ohm; Spiegel)*. It has already been described in *Spiegel and Sommer's* book (1944) that an optokinetic nystagmus may suppress a labyrinthine or spontaneous central nystagmus. Following lesions of the vestibular nuclei various interference phenomena between the spontaneous central and an experimental optokinetic nystagmus were observed (*Scala and Spiegel*, 1940). Recently several symposia have been held such as the oculomotor symposium, 1961, under the direction of M. *Bender. Adams* (1959) studied the relationship between postrotatory and optokinetic nystagmus and described it as follows "Die vestibulären Reaktionen sind beim Menschen den optokinetischen untergeordnet". In 1962 at the Physiological Congress in Leyden, a symposium was held, and optic and vestibular factors in motor coordination were discussed. *Szentágothai* presented a model of the optic-vestibular system and *Ter Braak* described that when an optokinetic and a vestibular stimulus act simultaneously, the resulting nystagmus is the algebraic sum of both expected nystagmus types; it is very possible that there is some common terminal center for both reflexes as has already been suggested by *Ohm* and by *Spiegel*. *Bender* concluded from his experiments with *Pasik* that integration between sensory (vestibular or visual) input and oculomotor output (nystagmus) must occur in widespread regions of the brain. *Jung* described that "Die Retikularis ist das zentrale Wahlorgan für die optisch vestibulären motorischen Koordinationen".

However, in the authors' view perrotatory and postrotatory as well as caloric nystagmus are qualitatively quite different from the standpoint of optic coordination. In order to emphasize this point, this article is reported.

Method

These experiments are conducted with the cooperation of several healthy adults.

Rotatory stimulation: Bárány's rotation technique is used. The subject is rotated 10 times in 20 seconds. Perrotatory and postrotatory nystagmus are recorded by ENG for detailed analysis.

Caloric stimulation: Caloric nystagmus is elicited by springing 20 cc of water at 20°C into one ear canal and is also recorded by ENG. The subject is placed supine on the bed with the head flexed 30 degrees so as to keep the horizontal semicircular canal in the vertical plane.

Optokinetic stimulation: A cylinder, one meter in diameter, is covered with white cloth on which are drawn 8 parallel stripes, 5 cm in width, with equal distances between them. This cylinder is placed at a 50 cm distance in front of the subject's eyes and rotated in the horizontal or vertical plane in order to induce an optokinetic response in horizontal or vertical direction. The cylinder is rotated at a constant speed of one rotation per 5 seconds in order to induce active and identical optokinetic nystagmus.

Results

Experiment 1: Influence of optokinetic stimulation on postrotatory nystagmus:

Rotation of a cylinder is started on shortly cessation of rotation on Bárány's chair, and the effect of optokinetic stimulation on postrotatory nystagmus is studied. Four different optokinetic stimuli are used: horizontal in the direction of postrotatory nystagmus or in the opposite direction; vertical upward and downward direction.

(a) Horizontal optokinetic stimulation. Same direction as postrotatory nystagmus.

Postrotatory nystagmus after rotation to right shows a small amplitude and relatively short duration. However, when optokinetic stimulation in the same direction is applied, the amplitude of the postrotatory nystagmus becomes greater.

(b) Horizontal optokinetic stimulation. Direction opposite to post-rotatory nystagmus.

When optokinetic nystagmus in a direction opposite to postrotatory nystagmus is produced immediately after the cessation of rotation, ENG shows a complicated pattern of response. Analysing this ENG, the following is observed (Fig. 1).

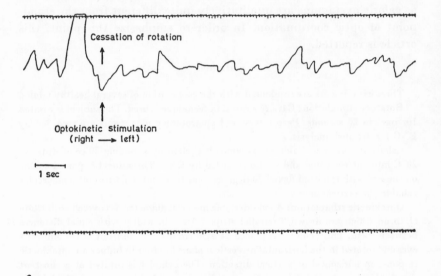

Fig. 1

First, two beats of very slow eye swings occur in the direction opposite to postrotatory nystagmus i.e. in the direction of optokinetic nystagmus, and a small pendulum like fluctuation follows. Then, an undulation of relatively greater amplitude appears and its direction (to the left) is opposite to the direction of the previous rotation on Bárány's chair. This leftward movement is the only postrotatory characteristic and following this, pendulum like fluctuation again appears. Finally active optokinetic nystagmus in a direction opposite to the postrotatory nystagmus is seen. These findings suggest that postrotatory nystagmus is completely suppressed by the optokinetic stimulation. As an arrow indicates in the lower part of Fig. 1, ENG shows originally true type of postrotatory nystagmus even though it is of slow speed after the cessation of rotation of a cylinder which gives the optokinetic stimulation opposite to the postrotatory nystagmus.

(c) Optokinetic stimulation in vertical direction.

Immediately after rightward *Bárány* rotation, upward vertical cylinder rotation is carried out. Fig. 2 shows records of the horizontal and vertical components of this induced nystagmus.

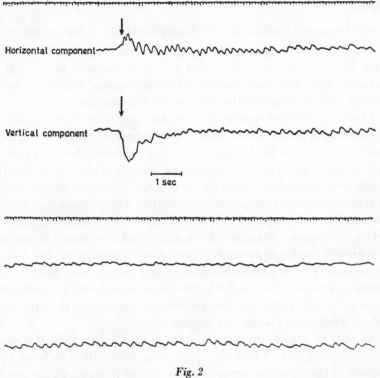

Horizontal component

Vertical component

1 sec

Fig. 2

Soon after the cessation of rotation, horizontal nystagmus with a vertical component is clearly recorded. During the first 3 seconds after cessation of *Bárány* rotation, postrotatory horizontal nystagmus appears to be more marked; however, 6 seconds after cessation of *Bárány* rotation, both horizontal and vertical components appear to be equally clear (with naked eye observation, eye balls move in a diagonal and move in the vector direction). Gradually, the horizontal component disappears and only the vertical component which is a response to optokinetic stimulation persists.

This finding suggests that movements of entirely different origin (horizontal postrotatory nystagmus and vertical optokinetic nystagmus) coordinate in one motion (in diagonal direction) at one stage, then finally postrotatory nystagmus is covered by optokinetic nystagmus which is the only type to persist. The anatomical and physiological mechanisms remain to be studied.

Experiment 2: Influence of optokinetic stimulation on caloric nystagmus:

(a) Horizontal optikinetic stimulation is added to caloric nystagmus:

Fig. 3 illustrates an experiment in which the right ear is syringed with 10 cc of water at 20°C and caloric nystagmus to the left is induced. After the occurrence of caloric nystagmus, horizontal optokinetic stimuli, either to the right or the left, are given.

First, the cylinder is rotated to produce optokinetic nystagmus in a direction opposite to the caloric nystagmus. At the moment indicated by the arrow in the upper part, the amplitude becomes greater and the direction of caloric nystagmus gradually changes to that of the optokinetic nystagmus. After the cessation of cylinder rotation, caloric nystagmus to the left reappears as indicated. This phenomenon clearly demonstrates the suppressing effect of optokinetic stimulation on the caloric nystagmus. 25 seconds later, the cylinder is rotated to produce optokinetic nystagmus which is the same direction as the caloric nystagmus. The amplitude becomes greater and the rate quicker. The caloric nystagmus is controlled by the characteristics of the optokinetic nystagmus. After stopping the optokinetic stimulation, the rhythm of caloric nystagmus resumes its original character.

As seen in Fig. 3, the caloric nystagmus is covered by the optokinetic nystagmus. When the optokinetic stimulation induces nystagmus in the direction opposite to the caloric nystagmus, the amplitude becomes greater but the speed becomes slower. When optokinetic stimulation is given in the same direction as the caloric nystagmus, the amplitude also becomes greater, with the speed faster. This is a very interesting phenomenon.

(b) Vertical optokinetic stimulation is added to the caloric nystagmus:

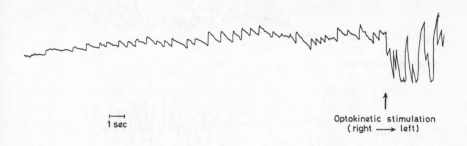

1 sec

Optokinetic stimulation
(right ⟶ left)

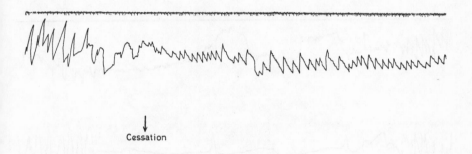

Cessation

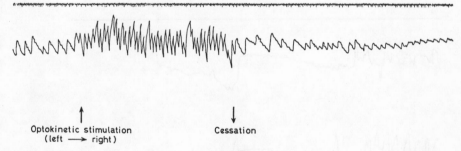

Optokinetic stimulation
(left ⟶ right)

Cessation

Fig. 3

In Fig. 4, 20 cc of water at 20°C is syringed into the right ear and caloric nystagmus is induced. The horizontal and vertical components are separately recorded.

During the presence of caloric nystagmus, a cylinder is rotated vertically upward. As soon as upward optokinetic stimulation is

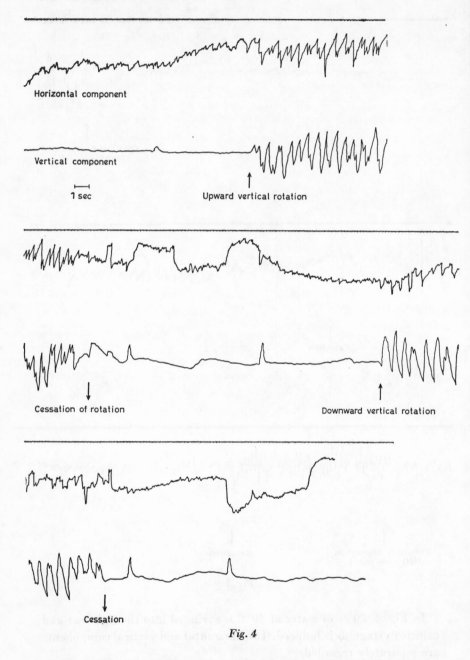

Horizontal component

Vertical component

1 sec

Upward vertical rotation

Cessation of rotation

Downward vertical rotation

Cessation

Fig. 4

given, the ENG shows clearly a downward nystagmus. It is interesting to note that the horizontal component, due to caloric stimulation, shows a larger amplitude and the speed of the eye movements coincides with that of the vertical component of optokinetic nystagmus. According to naked eye observation, the eye balls move diagonally. Greater amplitude is also visible. After the cessation of cylinder rotation, the horizontal component returns to the original type and the amplitude becomes smaller. 20 seconds later, the cylinder is rotated vertically downward in order to induce upward optokinetic nystagmus. The ENG shows a clear vertical component upward. In this case, the horizontal component increases and it coincides with the optokinetic nystagmus. ENG also shows increased eye speed. After cessation of cylinder rotation, the horizontal component returns to the previous pattern.

Summary

(1) Amplitude and speed of postrotatory nystagmus are increased by optokinetic nystagmus when the optokinetic stimulation is applied to induce the same direction of the quick component as the postrotatory nystagmus. Postrotatory nystagmus is suppressed by optokinetic nystagmus when the optokinetic stimulation is applied to induce the direction of the quick component opposite to that of the postrotatory nystagmus. When the optokinetic stimulation ceases, the original postrotatory nystagmus is demonstrable even more than 10 seconds after the cessation of rotation.

(2) When vertical optokinetic stimulation is added to the postrotatory nystagmus, the ENG shows not only a horizontal component but also a vertical component. The eye balls move diagonally with a certain rhythm. Gradually the horizontal component disappears and only the vertical component of optokinetic nystagmus persists. The two different rhythms of postrotatory and optokinetic nystagmus result in optokinetic rhythm.

(3) When the optokinetic stimulation is applied to induce a direction opposite to that of caloric nystagmus, the amplitude of caloric nystagmus becomes greater and the quick component of caloric nystagmus is changed to that of optokinetic nystagmus. After the cessation of optokinetic stimulation, caloric nystagmus resumes the previous smaller amplitude and character. When the optokinetic stimulation is given to induce the same direction as the caloric nystagmus, the direction of caloric nystagmus, of course, remains the same but the amplitude and eye speed increase. Again, on removal of this optokinetic stimulation, the caloric nystagmus resumes its original character.

(4) When vertical optokinetic stimulation, downward or upward, is added to the caloric nystagmus, ENG shows a vertical as well as horizontal component. The eye balls move diagonally. The horizontal component due to caloric stimulation shows a larger amplitude and the speed of eye movements coincides with that of the vertical component of optokinetic stimulation. After cessation of the cylinder rotation, the horizontal component returns to the previous pattern.

Zusammenfassung

1. Amplitude und Geschwindigkeit des postrotatorischen Nystagmus werden erhöht durch optokinetischen Nystagmus, wenn die optokinetische Reizung die gleiche Richtung der raschen Komponente erzeugt wie die des postrotatorischen Nystagmus. Postrotatorischer Nystagmus wird unterdrückt durch optokinetischen Nystagmus, wenn die optokinetische Reizung eine rasche Komponente entgegengesetzt dem postrotatorischen Nystagmus auslöst. Nach Beendung der optokinetischen Reizung ist der original postrotatorische Nystagmus noch länger als 10 Sekunden wahrnehmbar.

2. Das Elektronystagmogram zeigt nicht nur eine horizontale, sondern auch eine vertikale Komponente, wenn vertikale optokinetische Reizung zum postrotatorischen Nystagmus hinzugefügt wird. Die Augen bewegen sich diagonal in einem gewissen Rhythmus. Die horizontale Komponente verschwindet allmählich und nur die vertikale Komponente des optokinetischen Nystagmus bleibt übrig. Die beiden Rhythmen des postrotatorischen und optokinetischen Nystagmus ergeben den optokinetischen Rhythmus.

3. Wenn die optokinetische Reizung einen Nystagmus in der Gegenrichtung zum kalorischen Nystagmus erzeugt, wird die Amplitude des kalorischen Nystagmus vergrößert, und die schnelle Komponente des Nystagmus schlägt in der Richtung des optokinetischen Nystagmus. Nach Beendigung der optokinetischen Reizung hat der kalorische Nystagmus wieder die frühere Amplitude und Richtung. Wenn die optokinetische Reizung Nystagmus in der gleichen Richtung wie der kalorische Nystagmus erzeugt, bleibt die Richtung des kalorischen Nystagmus unverändert, aber die Amplitude und Schnelligkeit der Augenbewegungen nehmen zu. Bei Aufhören der optokinetischen Reizung nimmt der kalorische Nystagmus wieder den früheren Charakter an.

4. Wenn vertikale optokinetische Reizung abwärts oder aufwärts zum kalorischen Nystagmus hinzukommt, zeigt das Elektro-Nystagmogram eine vertikale sowie eine horizontale Komponente. Die Augen bewegen sich in diagonaler Richtung. Die durch kalorische Stimulation hervorgerufene horizontale Komponente zeigt eine höhere Amplitude und die Geschwindigkeit der Augenbewegungen deckt sich mit der der vertikalen Komponente der optischen Reizung. Die horizontale Komponente kehrt nach Beendigung der Zylinderrotation zur ursprünglichen Form zurück.

Résumé

1° L'amplitude et la vitesse du nystagmus post-rotatoire sont accrues par le nystagmus optocinétique lorsque la stimulation optocinétique est orientée dans la même direction que la composition rapide du nystagmus post-rotatoire. Le nystagmus post-rotatoire est supprimé par le nystagmus optocinétique lorsque la stimulation optocinétique est orientée en sens opposé. Lorsque la stimulation optocinétique cesse, on peut mettre en évidence le nystagmus post-rotatoire original pendant plus de 10 secondes après la cessation de la rotation.

2° Lorsqu'une stimulation optocinétique verticale est ajoutée au nystagmus post-rotatoire, l'ENG montre non seulement une composante horizontale, mais aussi une composante verticale. Les globes oculaires se déplacent en direction diagonale selon un certain rythme. La composante horizontale disparaît progressivement et seule persiste la composante verticale du nystagmus optocinétique. Les deux rythmes différents du nystagmus post-rotatoire et du nystagmus optocinétique donnent le rythme optocinétique.

3° Lorsque la stimulation optocinétique est appliquée en direction opposée au nystagmus calorique, l'amplitude du nystagmus calorique augmente et sa composante rapide est remplacée par celle du nystagmus optocinétique, le nystagmus calorique reprend son amplitude plus faible et son tracé antérieur. Lorsque la stimulation optocinétique s'exerce dans la même direction que le nystagmus calorique, la direction de ce dernier demeure évidemment la même, mais son amplitude et sa rapidité augmentent de nouveau. La suppression de la stimulation optocinétique rend au nystagmus calorique son caractère original.

4° Lorsqu'une stimulation optocinétique verticale, vers en bas ou vers en haut, est ajoutée au nystagmus calorique, l'ENG montre une composante verticale et une composante horizontale. Les globes oculaires se déplacent en direction diagonale. La composante horizontale due à la stimulation calorique présente une amplitude plus grande et la rapidité des mouvements oculaires coïncide avec celle de la composante verticale de la stimulation optocinétique. Après cessation de la rotation du cylindre, la composante horizontale retrouve son tracé antérieur.

Resumen

1.° La amplitud y velocidad del nistagmus rotativo se incrementa por la estimulación optoquinética si esta da lugar a un nistagmus que en el componente rápido tenga la misma dirección que el nistagmus rotatorio. El nistagmus rotatorio se inhibe por el contrario por el nistagmus optoquinético, si el componente rápido de est último tiene una dirección opuesta a la del nistagmus rotatorio; después de cesar la estimulación optoquinética puede persistir el nistagmus rotatorio original por un espacio de tiempo incluso superior a los 10 segundos.

2.° El electronistagmograma no solo muestra el componente horizontal sino tambien el vertical cuando se suma una estimulación optoquinética vertical al nistagmus rotatorio. Los ojos se mueven en un determinado ritmo en sentido diagonal desapareciendo gradualmente el componente horizontal y persistiendo solamente el componente vertical del nistagmus optoquinético. Los ritmos del nistagmus rotatorio y optoquinético resultan en el ritmo optoquinético.

3.° Cuando se provoca mediante estimulación optoquinética un nistagmus de dirección opuesta al nistagmus calórico, se incrementa la amplitud de este apareciendo el componente rápido en dirección al nistagmus optoquinético. Después de cesar la estimulación optoquinética el nistagmus calórico recobra su previa amplitud y dirección. Si la estimulación optoquinética desencadena un nistagmus en la misma dirección de este pero se incrementa la amplitud y velocidad de los ojos. Cuando cesa el estímulo optoquinético el nistagmus calórico recupera sus características iniciales.

4.° Si se suma una estimulación optoquinética vertical al nistagmus calórico el electronistagmograma mostrará tanto el componente vertical como el horizontal. Los ojos se mueven en dirección diagonal. El componente horizontal desencadenado por la estimulación calórica, incrementa su amplitud y la velocidad de los movimientos de los ojos coincide con la del componente vertical provocado por la estimulación optoquinética. Despues de cesar la rotación del cilindro el componente horizontal recupera sus características iniciales.

References

Adams, A.: Elektronystagmographische Untersuchungen über die optisch-vestibuläre Integration usw. Pflügers Arch. ges. Physiol. *269:* 344 (1959).

Bender, M. B.: O.K.N., V.N. and O.K.A.N. Proc. int. union physiol. sciences *1:* 508 (1962).

Fukuda, T.: The unidirectionality of the labyrinthine reflex etc. Acta oto-laryng., Stockh. *50:* 507 (1959).

Fukuda, T.; Hinoki, M. and *Tokita, T.:* Provocation of labyrinthine reflex by visual stimuli. Acta oto-laryng., Stockh. *48:* 425 (1957).

Fukuda, T. und *Tokita, T.:* Über die Beziehung der Richtung der optischen Reize usw. Acta oto-laryng., Stockh. *48:* 415 (1957).

Jung, R.: Zusammenfassung. Proc. int. union physiol. sciences *1:* 518 (1962).

Ohm, J.: Über den Einfluß des Sehens auf den vestibulären Drehnystagmus. Z. Hals-Nasen-Ohrenheilk. *16:* 521 (1926).

Scala, N. P. and *Spiegel, E. A.:* Subcortical (passive) optokinetic nystagmus in lesions of the midbrain and of the vestibular nuclei. Conf. Neurol. *3:* 53–73 (1940).

Spiegel, E. A. and *Sommer, I.:* Neurology of the eye, ear, nose and throat (Grune & Stratton, New York 1944).

Szentágothai, J.: Anatomical basis of visuo-vestibular coordination etc. Proc. int. union physiol. sciences *1:* 485 (1962).

Ter Braak, J. W. G.: Optokinetic center of eye movements etc. Proc. int. union physiol. sciences *1:* 502 (1962).

Authors' address; Prof. Dr. Tadashi Fukuda, Ent-Dept. Gifu Medical School, Tsukasamachi 40, *Gifu* (Japan)

Acta Otolaryng 71: 282–287, 1971

PHYSIOLOGY OF NYSTAGMUS

T. Fukuda and T. Tokita

From the E.N.T. Department of Gifu Medical School, Gifu, Japan

Abstract. We felt it necessary to record not only eye movements but also the movements of the head in order to study the physiology of nystagmus; therefore we constructed a telemeter to carry out these tasks simultaneously. By the use of this telemeter, we established that nystagmus is a very important motor reflex which can be observed in human actions and behaviour.

We would like to demonstrate some recent interesting findings on nystagmus during voluntary bodily movement as recorded by means of a radio-telemeter. The present telemetry system was designed to record not only eye movements but also movements of the head in order to investigate the relationship between the movements of these two parts of the body. Now we shall demonstrate our findings as revealed by the simultaneous graphic recording of eye and head movements.

As is shown in Fig. 1, electrodes are applied to the canthus regions on both sides of the head in order to record eye movements. Three accelerometers are attached to a helmet, one on the top, one on the right side and remaining one at the rear. Three accelerometers are able to record lateral, anterior posterior and vertical movements respectively. Channel 1 records horizontal eye movements. Channels 2, 3, 4 record the degree and direction of accelerations of the head in three axes, i.e. lateral, anterior posterior and vertical axes as shown in Fig. 2.

When a subject began to walk, spikes appeared in Channel 4, indicating vertical head movements and, on turning around or changing the direction of walking, brisk eye nystag-

mus occurred as is seen in Channel 1 of Fig. 3. However, when a subject undergoing drill changes direction on command, such multiple nystagmic movements do not occur, one change in direction being accompanied by only one nystagmus is clearly shown in Fig. 4.

Here you see a rotation which is called "pirouette" in ballet terminology. One rotation was accompanied by one nystagmus, and the nystagmus is not only a movement of the eyes but also a movement of the head. Two rotations of the same turn evoked two nystagmic movements as shown in Fig. 5. The eye movements are indicated in Channel 1 and the horizontal head movements in Channels 2, 3. It is important to note that two nystagmic movements of the eyes are accompanied by two nystagmic movements of the head, and the head movements precede eye movements.

Now, a ballet turn which is called "tour de chaines" is shown. This turn is performed by both a skilful ballet dancer and a novice. The former showed synchronous nystagmic movements of the eyes and head as is shown in Fig. 6, whereas the latter exhibited very erratic and irregular head and eye movements. It is important to notice the irregular vertical movements of the head which is shown in Channel 4.

Now, you still observe the eye movements of various subjects during a skating spin. The graphic record (Fig. 7) of a well trained skater showed many small but regular nystagmus during a spin. The graphic record of a moderately trained subject showed larger amplitude but

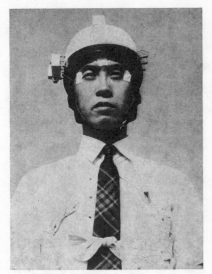

Fig. 1. Radiotelemeter to record both eye and head movements simultaneously.

Nystagmus on Changing the Direction of the Gait

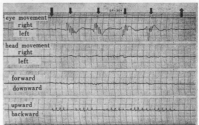

Fig. 3. Brisk nystagmic movements of the eye on turning the direction of walking in channel 1.

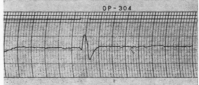

Fig. 4. One change in direction undergoing drill was accompanied by only one nystagmus.

less frequent nystagmus, whereas the record of a novice showed very irregular and erratic nystagmus. It is very interesting to note that marked eye nystagmus was clearly evidenced during rapid turnings of the spin.

Next, we would like to show the nystagmus of a skilful ballet dancer during and after turns

in which he was instructed to keep his eyes open and then closed. As you see in Fig. 8, when his eyes remained open, no eye movement could be observed after he finished the turn, in other words, no postrotatory nystag-

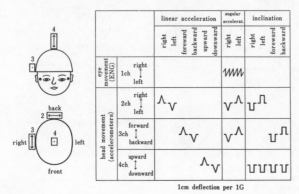

1cm deflection per 1G

Fig. 2. Records of 4 channels.

mus occurred. Then he was asked to close his eyes, after finishing the turn. Many beats of nystagmus were recorded for a long time after closing his eyes. This record is in clear contrast to the one with his eyes open and this fact was reported by Dr Collins and others (Collins, 1966).

Fig. 9 shows some interesting findings on nystagmus of the same dancer after he had been rotated passively 10 times in 20 sec in a rotating chair, instead of turning actively through his own efforts. His eyes were open. After the rotation, his eyes showed no nystagmic movements. Here you can see brisk eye nystagmus occurred during the rotation, but once the rotation ceased, eye movement was scarcely detectable as you see in the recording. When a physically untrained person (a normal adult) was rotated in an identical manner, a marked postrotatory nystagmus could be seen. Another characteristic phenomenon observed in the skilful ballet dancer is single large nys-

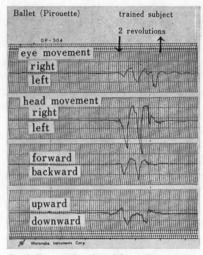

Fig. 5. Two rotations in a "pirouette" were accompanied by two nystagmic movements of the eye as well as of the head.

Ballet (Tours chaînés)

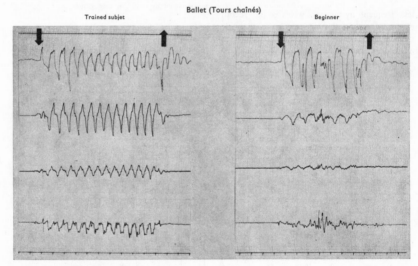

Fig. 6. Nystagmic movements of the eye and of the head caused by rotations of a "tour de chaines" of a trained subject and of a beginner.

Skating (Spin)

Trained subject

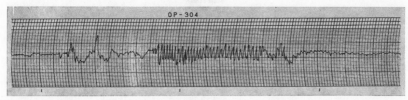

Moderately trained subject

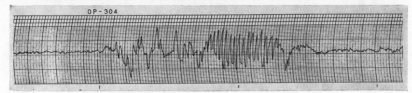

Beginne

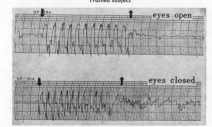

Fig. 7. Graphic records of eye movements during a skating spin.

Ballet (Fouetté en tournant)

Trained subject

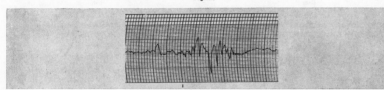

Fig. 8. Records of nystagmus during and after active rotations of a skilful ballet dancer with eyes closed and then open.

tagmus accompanied by each rotation. Ten nystagmus movements were observed during the ten rotations. With a control subject, nystagmus of inconsistent amplitude occurred irregularly during rotation in a rotating chair.

Finally we would like to mention some differences between a skating spin and a ballet turn. Here you can compare a "tour de chaines" with a skating spin in Fig. 10. In the former, the eyes and head show unison-coincident movements. In the latter, the eyes showed brisk nystagmic movements whereas the insignificant head movements were neither nystagmic nor rhythmic.

Passive Rotation

Ballet dancer

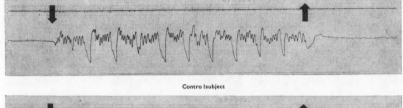

Contro Isubject

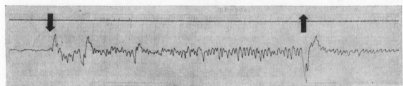

Fig. 9. Records of nystagmus during and after passive rotations on a rotating chair of the same skilful dancer and of a control subject.

Ballet   Skating

Tours chaînás   Spin

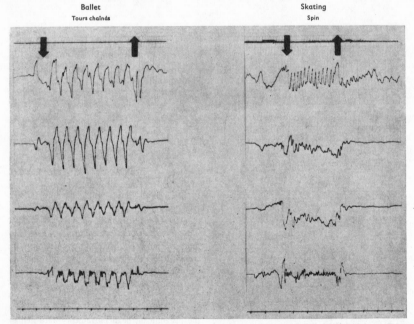

Fig. 10. Records of eye and head movements of a "tour de chaines" in comparison with records of a skating spin.

Acta Otolaryng 71

Thus it is concluded that the nystagmus is an essential physiological factor of our daily locomotion. Human beings have learned to control this reflex phenomenon "nystagmus" ingeniously and to utilize it for the smooth performance of our bodily movement: therefore, nystagmus is one of the most important and essential motor reflexes which can be observed in human daily locomotion and behaviour.

ACKNOWLEDGMENT

This paper is a synthesis of a large number of experimental studies. The investigations were carried out by Dr T. Tokita M.D., with the collaboration of Dr T. Watanabe M.D., Dr M. Ogushi, Dr H. Miyata M.D., Dr M. Fujigaki M.D., Dr T. Kobayashi, Dr T. Nagata, Dr K. Kato M.D., Dr Y. Kato, Dr T. Hibi, Dr R. Shimada, Dr T. Suzuki, Dr T. Taguchi and Dr Y. Hayano, all of them are members of the E.N.T. Department of the Gifu Medical School.

REFERENCES

Battye, C. K. & Joseph, J. 1966. An investigation by telemetering of the activity of some muscles in walking. Med Biol Engng 4, 125.

Collins, W. E. 1966. Vestibular responses from figure skaters. Aerospace Med 37, 1098.

McCabe, B. F. 1960. Vestibular suppression in figure skaters. Trans Amer Acad Ophthal Otolaryng 64, 264.

Osterhammel, P. & Peiterson, E. 1968. Telemetry system for nystagmus. Acta Otolaryng (Stockh.) 65, 527.

Osterhammel, P., Terkildsen, K. & Zilstorff, K. 1968. Vestibular responses following high velocity rotation. Acta Otolaryng (Stockh.) 66, 145.

T. Fukuda, M.D.
E.N.T. Department
Gifu Medical School
Tsukasamachi 40
Gifu
Japan

DISCUSSION

R. Hinchcliffe: Has Mr Fukuda employed this method in the investigation of patients complaining of vertigo? The potentialities are obviously great.

J. D. Hood: Mr Fukuda is to be congratulated upon his excellent film which is both stimulating and instructive. Although it raises many issues there is one to which I should like to draw particular attention namely the demonstration that following upon the cessation of a pirouette a ballet dancer develops a brisk nystagmus with closed eyes. Similar results to these have also been reported by Collins in the case of ice skaters. There is however an essential difference between the two because whereas an ice skater invariably keeps his head in rigid conformity with his body during a spin, a ballet dancer in the course of a pirouette fixes his gaze upon some stationary object and each complete rotation of the body is followed by a very rapid 360° turn of the head when the object is refixated. It is commonly held that this manoeuvre eliminates the post-rotatory stimulus that would otherwise result and consequently accounts for the absence of vertigo. I myself have always questioned this argument because if it were true then it is difficult to account for the fact that ballet dancers habituate since they would never be subject to a habituating stimulus.

Mr Fukuda's excellent film now shows quite clearly that a post-rotatory stimulus does exist despite the rapid rotation of the head and it now seems clear that this is the source of the habituating stimulus.

T. Fukuda (Reply) to Mr *Hinchcliffe:* Cordial thanks for your suggestion. The eye and head movements of "patients suffering from vertigo" are not yet investigated. I hope to have a chance to report the results obtained in the near future.

Acta Otolaryngol Suppl 330: 9–14, 1975

POSTURAL BEHAVIOR AND MOTION SICKNESS[1]

Tadashi Fukuda

From the Department of Otorhinolaryngology, Gifu University School of Medicine, Gifu, Japan

Abstract. (1) An interesting style of acting was demonstrated on a stage of a Kyogen, a classic comedy of Japan, titled "Funawatashimuko", i.e., "A boatman and a bridegroom in a boat". Antagonistic postures which move toward the opposite direction were displayed by a boatman who is pulling an oar and a passenger who is being moved by the rolling of a boat. (2) Why doesn't one suffer from motion sickness when he drives a car but may suffer from it when he is a passenger? This question was answered, from the standpoint of human postures, by observing the antagonistic postures exhibited by a bus-driver and a passenger, and also by the findings in postrotatory eye nystagmus (an indication of artificial motion sickness) which was varied according to the three different positions of the head. (3) Learning postural adjustment against motion sickness, developing through repetitively traveling in vehicles, was also objectively shown in a posture of an experienced lady bus-conductor whose body inclined in the same direction as that of the driver. Its similarity to the establishment of a kinetic labyrinthine reflex in chickens was explained.

Ladies and Gentlemen!

Several years ago, I happened to be watching a very interesting Kyogen (a very old traditional comedy of Japan) entitled "Funawatashimuko", i.e., "A bridegroom and a boatmen in a boat". The comedy was very delightful and it was shown on a New Year's television program. At that time, I had been working on the problem concerning the relation between sea sickness and human postures, and I was deeply impressed by the style in which the actors portrayed the bodily movements of men in a boat, a style of acting which has remained unchanged for 500 years. In order to explain more clearly the relationship between motion sickness and human

posture, a demonstration will be given on this stage at the opening ceremony. The Shigeyamas, the head family of Kyogen in Kyoto, have accepted to perform this comedy, much to my pleasure. Before the curtain is raised, I would like to emphasize a few important points. Very briefly, the story behind this Kyogen goes as follows. A bridegroom is going to visit the parents of his bride and must cross the Biwako Lake in a small boat (like a Venetian gondola). There is a boatman with an oar in his hand waiting on the shore for passengers on his row boat.

A bridegroom who is carrying with him a sealed keg of Sake to be presented to the bride's father, hires this boatman. Because the weather is very cold and windy, the boatman who is an admirer of Sake (or an elbow bender), repeatedly asks for a cup of Sake. Of course, the bridegroom ignores the request. The boatman is very thirsty and refuses to continue rowing until finally the disgusted bridegroom gives him a cup of Sake by unsealing the keg. As the play continues, the bridegroom who is also an elbow bender becomes thirsty and drinks up all the Sake from the keg with the rather annoyed boatman. The important point which I wish to show and emphasize in this play is each posture of the boatman who is rowing and of the passenger on the boat being moved by rowing. Now, let me show you the point by figures!

As you see in Fig. 1, the boatman stretches his upper extremities to push the oar, bending forward, his posture inclines to your left. In contrast, the bridegroom inclines in the opposite direction, to your right. Thus the direction

[1] This paper was read at the opening ceremony of the 73rd Congress of Japan Society of Otorhinolaryngology held in Gifu City 1972 by the author who was the President.

Fig. 1. The boatman, who is standing, stretches his upper extremities to an oar, bending forward and his posture inclines toward your left. On the contrary, the sitting bridegroom inclines in the opposite direction, being moved by the roll and his posture inclines toward your right. Thus the directions of postural inclination in a boat are opposite to each other.

been handed down generation after generation. Now, let us consider the movement of a boat as played in the theatrical performance shown in Figs. 1 and 2. The boat is moving straight from the back of the stage toward the audience. This movement of the boat is made possible by the work of a boatman who pushes and pulls an oar alternately with a delicate twist of the wrists. As the boat moves forward, it also undergoes a partial rotation (roll) along its longitudinal axis. It must also be noted that the movement of body of the boatman is in the same phase as the longitudinal rotation (roll) of the boat and that of the passenger is in the opposite direction to the rolling movement of the boat. Thus the behavior of the movement of two men in a boat is played in this example from the theatrical performance of a Kyogen play. Recently, I had a chance to see the staging of one of Japan's modern dances which portrayed the actions of a boatman and a passenger in the same boat. In this play, the boatman who was

of inclination of the two postures in the boat are opposite each other. As the rowing action continues, the inclination of the posture of the two individuals changes direction as shown in Fig. 2. The boatman who is pulling the oar bends backward, i.e., the posture inclines to your right, while the posture of the bridegroom inclines toward your left. This alternating pattern of individual posture continues changing rhythmically as the rowing continues. When, on the stage, the boatman and the bridegroom move in opposite directions rhythmically, the audiences feel just as if they saw a boatman and a passenger in a real boat hurrying its way through the waves of the Biwako Lake. The reason why such a realistic impression is imparted by this highly stylized performance of the play can be found in the style of performance in which the postures of the two men incline to the opposite direction alternately, a style which has probably

Fig. 2. As the rowing action continues, the direction of postural inclination changes to the opposite sides. The head and trunk of the boatman who is pulling the oar bends backward, i.e., the posture inclines toward your right, while the bridegroom inclines toward your left. Thus the two postures move into opposite directions.

rowing the boat and the passenger who was being moved by the rolling of the boat, swayed in the same direction, i.e., both moved to your right or left at the same time. Looking at this modern dance, I was impressed by the accuracy with which our ancestors had translated a segment of real-life behavior to the artificial settings of the stage.

Similar contrasts in bodily postures, such as that just illustrated for a boatman and a passenger, can be observed in our daily life. For example, as shown in Fig. 3, the head and trunk of a driver of a bus inclines toward the right when he revolves the handle to his right in order to turn the bus to the right, while the head and trunk of a passenger in the same bus inclines toward the left. Thus, the postures of the two individuals incline in opposite directions. From the viewpoint of centrifugal force, it is interesting to note that the driver deviates across the midline against the centrifugal force, showing a centripetal posture, while the passenger deviates across the midline in the direction opposite to the rotation of the bus, showing a centrifugal posture. Thus opposite postures are clearly evidenced between a driver and passenger in a bus. It is very important to remember that these contrasting postures are not the products of actors on a stage, but are real physiological phenomena caused by the movement of a bus which curves to the right.

The author would like to emphasize that such antagonistic patterns of postures as seen with a driver and a passenger in a moving bus as well as a boatman and a passenger (bridegroom) in a boat on a symbolic stage of Kyogen is deeply related to the cause of motion sickness. A driver as well as a boatman does not normally suffer from motion sickness while a passenger often times does. Moreover, we hear that one does not get motion sickness when he drives a car but when he is in a car as a passenger he does get it. Why is there such a difference concerning motion sickness between a driver and a passenger even though they are in the same car and subject to the same movement? The author would like to demonstrate that the difference is

Fig. 3. The head and trunk of the driver of a bus incline toward your right when he revolves the handle in a clockwise direction to make a right turn, while the head and trunk of a passenger in the same bus incline toward your left. Thus the two postures incline toward opposite directions from each other.

due to the difference in the postures in which the bodies incline in the opposite directions and cross each other, by introducing the following observation and experiment.

Güttich (1940) had observed that the position of the eyes and head of a human being is antagonistic in active and passive rotations. When a person begins to rotate around his own long axis, the eyes and the head deviate in the direction of rotation over the midline of the body and continue in this position during the period of the rotation accompanied by eye nystagmus. That is, an active rotation of a person is preceded by the eyes and head which deviate in the direction of rotation. However, when a person on a chair is rotated passively, the eyes and head deviate against the direction of rotation over the midline of the body and continue in this position accompanied by eye nystagmus during the period of the rotation, i.e., the eyes and head remain in the direction opposite the rotation of a chair during a passive rotation. Thus, with active rotation, the position of the eyes and head deviates in the direction of the rotation, while, with a passive rotation, it deviates against the direction of the rotation of a chair. In this way each position assumes an antagonistic, symmetric posture over the midline of the body.

This antagonism is observed not only during rotatory movement around one's own axis but also in other linear and circular movements and it stands as a general rule of postures differentiating active and passive movements of human beings. For example, when a runner on a track comes to a curve, his posture inclines toward the inside of the curve and he assumes a centripetal posture. However, as seen in Fig. 3, a passenger who is every inch passively moved in a bus assumes a posture which deviates to the outside of a curve due to the action of the centrifugal force. Thus the antagonism of the posture is also evidenced between a runner and a passenger, i.e., between active and passive movements at a curve in human.

It is of interest to examine the posture of the driver in Fig. 3 from the standpoint of the Güttich's observation. The driver does not assume a centrifugal posture such as seen in the passenger; the driver assumes the same centripetal posture as a runner going around a curve. Although the passenger and driver are in the same bus, the position of the head and trunk of the former deviates in a direction opposite to the curve, while that of the latter deviates in the same direction. The two postures incline in opposite directions from each other and assume antagonistic positions. Now, returning to why a person when driving a car does not suffer from motion sickness while he may do so when he is a passenger in a car. To answer this question, the author hypothesized that this is due to the contrast in assumed postures between drivers and passengers. To test this hypothesis the following investigations were performed.

Postrotatory nystagmus after 10 rotations in 20 seconds was adopted as an indicator of motion sickness, based on the following theory. Post-

rotatory nystagmus is a phenomenon arising from transient artificial labyrinthine ataxia caused by repeated rotations with which human beings are not familiar. Postrotatory sensation is that of vertigo or illusion, i.e., a transient motion sickness in which one feels as if still being rotated after the end of the rotation. In other words, the rotation test is a method for measuring the time required by a subject on whom unexperienced passive repeated rotations are imposed, to regain almost normal equilibrium of the body after transient labyrinthine ataxia. Postrotatory nystagmus and sensation may be observed as the symptoms of artificial motion sickness which are often accompanied by nausea. So, the recorded duration and the number of beats of postrotatory eye nystagmus, serve as an objective index of artificial motion sickness and its magnitude.

The chair was designed to turn in a complete circle, in clockwise as well as counter clockwise horizontal direction, and constructed to allow a subject's head to be held in different positions. Seventy healthy adults, male and female, without any ear diseases were rotated 10 times in 20 seconds in a counter clockwise direction. During rotation, the eyes were closed and the head was fixed in 3 different positions as follows: first head position: normal, i.e., 30° prone (nose down) position from the German horizontal plane, second head position: extreme left, i.e., toward the direction of chair rotation; and third head position: extreme right, i.e., against the rotation of chair. In each of the positions, the head was firmly fixed with a band to the head holder attached to the turning chair and the subject was rotated three sequences with 5 minute intervals. Immediately after each rotation, the eyes were opened and the examiner calculated the duration and beats of postrotatory nystagmus which showed marked decrease and increase according to the difference of the head position as shown in Fig. 4. In the case of the first head position, i.e., the normal head position, the numerical mean value of postrotatory nystagmus for 70 subjects was 27 beats in 21 seconds. With the head positioned in extreme left (the second head position), it was 21 beats in 16 seconds; and in the third position, against the direction of chair rotation, it was 39 beats in 25 seconds. To summarize, by fixing the head toward the direction of chair rotation (the second head position) the mean value of duration and beats of postrotatory nystagmus decreased markedly compared with the effect of fixing the head in normal head position (the first head position). In contrast, by fixing the head against the direction of chair rotation (the third rotation), it increased markedly.

It is very interesting to examine the results from the viewpoint of Güttich's observation described above that during active rotation the position of the eyes and head deviates toward the direction of rotation, while, during passive rotation, it deviates against the direction of rotation. In the present experiment, each rotation was performed passively by turning a chair. However, the head was fixed in three different positions based on the following assumption. The first head position, which is the normal position of old Barany rotation, was adopted as the standard position. In the second rotation, the head of a passively rotated subject was fixed in the same position as in a subject who turns

Fig. 5. A posture of an experienced lady bus-conductor while the bus is making a right turn. Even being moved passively in a bus, her posture (head and trunk) inclined toward the right, i.e., in the same direction as that of the driver, and opposite to that of a passenger; shown in Fig. 3.

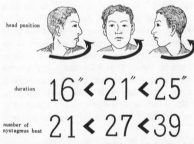

head position

duration $16'' < 21'' < 25''$

number of nystagmus beat $21 < 27 < 39$

Fig. 4. With head rotated and fixed toward the direction of chair rotation (the second head position), the mean value of duration in seconds and the number of beats of postrotatory nystagmus decreased markedly compared with the case in which the head was fixed in straight head position (the first head position). On the contrary, with head rotated and fixed opposite the direction of chair rotation (the third rotation), it increased markedly. Arrow: direction of chair rotation.

actively along his own axis. In the third rotation, the head was fixed against the direction of rotation. This is the head position of a subject during passive rotation in a chair in whom the head `is not fixed during rotation. For the reason described above, the author would like to call the second head position, the active position of the head, and the third position, the passive position.

As shown in Fig. 3, the head and trunk of a driver incline toward the right, i.e., toward the direction of the center of the curve and assume a centripetal posture. This posture may be said to correspond with the second head position, i.e., active position of the head resistant to postrotatory nystagmus, namely a posture in which a subject rarely suffers from motion sickness. On the contrary, the head and trunk of a passenger incline toward the left according to centrifugal forces and assumes a posture which is opposite the posture of a driver. The passenger's posture may be said to correspond with the third head position, i.e., passive position of the head

Acta Otolaryngol Suppl 330

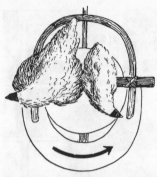

Fig. 6. Trained (right) and untrained (left) chickins. 10 seconds after the beginning of rotation. The head of the trained turns in the same direction of rotation through the kinetic labyrinthine reflex, while the head of the untrained deviates still in the opposite direction of rotation by the static labyrinthine reflex (normal deviation).

vulnerable to postrotatory nystagmus, that is a posture in which one easily develops motion sickness. Thus the marked decrease and increase of postrotatory nystagmus observed in the above investigation clearly explains the fact that a subject who is driving a car hardly ever suffers from motion sickness while he easily gets it when he is a passenger in a car.

Before summarizing the results, the author would like to add a few comments on the problem of learned postural adjustment or habituation. It is widely known that sailors or busconductors who have accumulated many years of experience in traveling on vehicles do not suffer from motion sickness. These people have developed special postural habits an example of which is illustrated in Fig. 5. The author observed the posture of a trained lady bus-conductor who had been working in a bus for 5 years and made photographs of her back (without her knowledge) while the bus was turning. When the bus is turned to the right, she assumed a posture with head and trunk inclined toward the right, i.e., exactly the same direction as the inclination of the driver. While being moved passively in the bus, her posture inclined in a direction opposite to that of the passenger shown in Fig.

3. It is very interesting to note that the passenger and lady bus-conductor while sitting on seats in the same bus, took exactly opposite postures and inclined in an opposite direction from each other. The lady bus-conductor took a posture similar to that of the driver whose posture was shown to be resistant to motion sickness. Thus a learned postural behavior or habituation is evidenced objectively in a lady bus-conductor who assumed an actively inclined posture toward the center of a curve, i.e., centripetal posture.

This centripetal posture was reported by the author about 20 years ago under the title of "Static and Kinetic Labyrinthine Reflex. Functional development of labyrinthine function with rotatory training" (Fukuda, 1958). To summarize the principal points shown in Fig. 6, blindfolded leghorns were rotated 100 times in 200 seconds to both directions every day for two weeks. It was observed after repeated rotations that the birds, during rotation, turned their heads in the direction of rotation. This phenomenon was never found in animals before repeated exposures to rotations. The labyrinthine function which caused it has been named the "Kinetic labyrinthine reflex". In contrast to this terminology, the deviation of the head against the direction of rotation has been called the "Static labyrinthine function", a reflex which was called "the normal deviation" to maintain the original position of the head. The animal showed a functional progress in equilibrating function through the repeated exposures to rotation, which were therefore termed training (Fukuda, 1963). Thus a learned postural behavior during rotation was evidenced objectively in the establishment of a kinetic labyrinthine reflex which is here again clearly observed in the posture of a trained lady bus-conductor.

REFERENCES

Güttich, A. 1940. Über den Antagonismus der Hals- und Bogengangs-reflexe. *Arch Ohrenheilk 147*, 1.

Fukuda, T. 1958. Static and kinetic labyrinthine reflex. *Acta otolaryngol 49*, 467.

Fukuda, T. 1963. Physiology of training. *Acta Otolaryngol 56*, 239.

Appendix B

The Reflex Physiology of Dynamic Postures

1. Introduction

In 1961, I wrote a paper in *Acta Oto-Laryngologica Supplement 161* entitled "Studies on Human Dynamic Postures from the Viewpoint of Postural Reflexes," which developed into Chapter 1 of this book. This paper attracted attention outside Japan, and I have been asked for permission to use some of the illustrations for other publications. As I readily grant this permission, my view has been quoted and introduced in publications abroad (e.g., Brookhart, 1979; Keele, in press).

After the publication of this supplement, I did not write any more about Magnus's postural reflex or the neck reflex in Japan. But in 1979, I gave an opening speech at the Symposium on Posture held in Amsterdam (Fukuda, 1979); on the occasion of the publication of this revised edition,* I would like to describe further developments in research in this field and evaluate postures from the standpoint of the postural reflex.

First, I will briefly describe my research on the neck reflex. In 1924, Magnus published *Körperstellung*, in which he recorded in detail his experiments on the postural reflex and demonstrated clinical examples. In this book he delineated the following points in relation to the neck reflex and the labyrinthine reflex. These reflexes, according to him, are postural reflexes typically seen in decerebrated animals, but on careful inspection, they are also clearly observable in daily postures of normal animals. These reflexes constitute the basis of movement. In animals higher than monkeys, these reflexes are difficult to observe. Particularly in humans, they are clearly discernible only as pathological reflexes in decerebration states caused by lesions such as gunshot wounds of the brain or cerebral hemorrhage. Magnus took a pessimistic view of relating normal human postures and movements clearly with the above-mentioned reflexes. He

---

* The second edition of the Japanese-language version of this book was published in 1981, after the first edition had been continually in print for 24 years.

stated that human postures and movements might be controlled by different principles, because humans walk erect on two legs. His view is quoted from *Körperstellung* in the following passage:

> Tonische Hals- und Labyrinthreflexe sind im gewöhnlichen Leben bei gesundenen Erwachsenen nicht zu sehen. Diese Reflexe sind auch beim normalen Affen meistens nicht zu beobachten. Beim erwachsenen Menschen mit dem Gewinnen des aufrechten Ganges müssen andere Gesetzmässigkeiten auftreten.

> (The tonic neck and labyrinthine reflexes cannot be seen in the ordinary life of healthy adults. These reflexes are not seen in normal monkeys. In grownups who can walk erect, different principles must apply.)

I studied various dynamic postures and demonstrated that Magnus's tonic neck reflex clearly formed the basis of normal human posture and facilitated maximum, efficient muscular performance. Figure 1.1 (p. 5) is a typical neck reflex seen in a child with cerebral palsy, a decerebrated state. With the head turned to the right, the upper and the lower limbs on the right side, or the side toward which the head is turned, are extended, whereas the upper and the lower limbs on the opposite side are flexed. The limb on the side toward which the head is turned is termed the jaw limb, and that on the opposite side is termed the skull limb. This neck reflex, in which the jaw limbs are extended and the skull limbs are flexed, was regarded exclusively as a clinical finding of brain disturbance comparable to decerebration. In Figure 1.6 (p. 9), I showed that the neck reflex acted in construction of the ball-catching posture in such a way that maximum muscular strength could be displayed. With the head turned to the left, the jaw limbs, both the left upper and lower limbs, are maximally extended to catch the flying ball. The skull limbs are both flexed fully, although the player himself is completely unaware of this. That is, this is a reflex movement. When one draws a vertical line on the picture of the player in this posture, one finds that the movement of the limbs on the right side of this line is performed by the player—a volitional movement of stretching himself to catch the ball. However, the flexing of his right upper and lower limbs, which are situated on the left of this line, is a movement of which the player is totally unconscious—an involuntary reflex movement. The ball is in the air above him and to his left side. The player stretches himself with the left upper limb extended high, a voluntary movement. Simultaneously, to the left of the line, the right upper and lower limbs are flexed to the fullest degree reflexly with the right upper limb flexed anteriorly and the right lower limb flexed posteriorly. The anteroposterior balance at this instant is well maintained and thus the ball-catching posture is assumed.

Figure 1.9 (p. 15) shows the heading posture of soccer. A line is drawn through the neck. Above this line, the head is turned to the left to hit the ball in a conscious and voluntary movement. Below the line, the left upper and lower limbs are fully extended, while the right upper and lower limbs are maximally flexed. All these movements are performed without the player's awareness; i.e., they are reflex movements brought about by the neck reflex. In Chapter 1, I have shown that in various athletic postures, muscle forces are fully manifested owing to a variety of postural reflex, including the neck reflex, which participates in seemingly entirely volitional movements of the body. Additionally, I have shown that the keen observation and intuitive understanding of the neck reflex has been revealed in the work of some artists as in the statue of a Buddhist Deva king of Kōfukuji (Fig. 1.35; p. 32) and the god of thunder drawn by Tawaraya Sōtatsu (Fig. 1.36; p. 33).

The above is an outline of my research. In the following section, later developments in research will be presented.

2. Bow-Drawing Posture, a Posture Which Conforms with the Neck Reflex

Nishihata (1938) first noted that the extended upper limb on the frontal side and the flexed upper limb on the occipital side when drawing a bow are results of head rotation, or the neck reflex. He performed research on the postural reflex in the region and made otological reference to the bow-drawing posture and other dynamic postures for the first time from the standpoint of reflex physiology in his book *Jibi-inkō-kagaku-sōron* [An Introduction to Otorhinopharyngolaryngology]. I would like to refer below to various postures, including the bow-drawing posture in which maximal muscle force is manifested in conformation with the neck reflex.

The statue of *Hercules Drawing a Bow* by Bourdelle (Fig. App. 1): The upper limb on the frontal side is correctly extended and the upper limb on the occipital side is flexed.

The statue of *Katsujinsen* by Denchū Hiragushi (Fig. App. 2): This also shows a typical posture conforming with the neck reflex. The extension of the limb on the frontal side is particularly striking.

God-General from the North (Fig. App. 3; a tricolored ceramic work from the Tang Dynasty, Kurashiki Museum): The extension and flexion brought about by the neck reflex are depicted not only in the upper limbs but also in the lower limbs and in the fingers.

Baseball pitching (Fig. App. 4): This is a difficult pitching posture of a baseball pitcher. It depicts the instant the ball leaves the fingers, when

FIG. APP. 1

FIG. APP. 3

FIG. APP. 2

Fig. App. 4   Fig. App. 5

Fig. App. 6

the muscle force is maximal. On analysis, one finds that the upper and lower limbs on the right side towards which the head is turned are fully extended, whereas the upper and lower limbs on the left side are flexed. This is a characteristic manifestation of the neck reflex in pitching a ball with the fullest force.

High jump (Fig. App. 5): In this picture, the head is rotated to the right and the upper and lower limbs on the right, occipital side are maximally extended. At this moment, the jumper is concentrating most, flexing the left lower limb and extending the right lower limb so that he can jump over the bar without touching it. As his posture complies with the neck reflex, this movement is achieved.

Statue of a fighter by Bourdelle (Fig. App. 6): In this work, the head is turned towards the left and the left upper limb is extended maximally with the fingers also stretched; the right upper limb, which is on the occipital side, is flexed and is about to strike down the sword. This is also consistent with the neck reflex.

3. Postures That Do Not Conform with the Neck Reflex

Picture of Christ: In Figure App. 7, Christ has his head turned to the left. The right upper limb, which is on the occipital side, is extended, and the left upper limb, which is on the frontal side, is flexed. According to the principle of the neck reflex, the skull limb should be flexed and the jaw limb extended. Why is this posture different?

Both humans and animals assume a certain position of the head in relation to the body trunk, which is called the normal head position. When the head is turned from this normal head position around the long or sagittal axis of the body, the upper and lower limbs are extended or flexed according to a certain form, which is consistent with the neck reflex. The change in deep sensation of neck muscles, which is brought about by head rotation, causes flexion and extension of the upper and lower limbs. The fact that the extension and flexion of groups of muscles give rise to extension and flexion of different groups of muscles is fundamental to muscle physiology. This phenomenon was once called *induzierte Tonus Veränderung* by Goldstein (1923).

In this picture of Christ, the right upper limb is extended and the left upper limb is bent at the elbow. The head is turned to the left, which is the opposite to that expected from the neck reflex. In the next scene, Christ will turn his head from the left, the side of the young rich man, to the right to look at the poor couple standing in the corner, which will agree with the neck reflex. In this picture of Christ, the extension of the right upper limb and flexion of the left upper limb affect the neck muscles in such a way that the head is turned toward the right, which is the opposite to what is expected for the normal neck reflex. This gives the impression that Christ is about to turn his head to the right at any moment. In contrast, the young rich man is depicted as being quite unwilling. His head is turned to

FIG. APP. 7

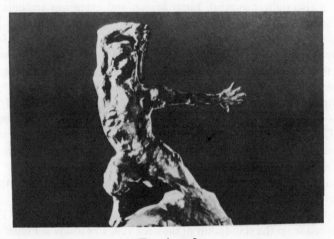

FIG. APP. 8

the left and his eyes are taken away from the couple to whom Christ points. The frontal upper limb of the young rich man is bent at the elbow and the fingers are also bent, resting on his waist. The occipital upper limb, i.e. the right limb, is stretched together with the fingers touching the desk. It seems that the young rich man does not want to look at the couple and wishes to go back, turning the head to the right, which the extension and flexion of the upper and lower limbs suggest. This posture gives the impression of unwillingness. The heads and the upper limbs of the two people are symmetrical, and they are facing each other. Their left upper limbs are both flexed, whereas the right upper limbs are both extended. These postures make Christ appear quite active and the young rich man passive.

Some statues by Bourdelle demonstrate characteristic postures of the neck reflex, as represented by *Hercules Drawing a Bow* (Fig. App. 1). The upper limb on the frontal side is extended and that on the occipital side is maximally flexed. This statue is an ideal model for explanation of the neck reflex. Bourdelle made many other statues of soldiers. Figure App. 6 shows a soldier with his head turned to the left. The left, or jaw, upper limb and even the fingers are maximally stretched, and the right, or skull, upper limb is flexed as the soldier wields a sword. This statue is very powerful, and one almost feels the breathing of the soldier because the posture completely agrees with the neck reflex. Figure App. 8 depicts another soldier, whose head is turned to the right, but the upper and lower limbs on the frontal side, which are to be extended according to the principle of the neck reflex, are flexed and the upper and lower limbs on the left are fully stretched, which is also contrary to the neck reflex. This statue is also quite powerful. Limb flexion and extension are opposite to those of the neck reflex and yet the posture is quite forceful. This is difficult to explain. My personal interpretation of this situation is as follows. In olden days on the battlefield, fighting took place not just one-on-one but between one soldier and several enemies. The soldier in Figure App. 8 may be being attacked by two or more enemies from different sides. The right upper limb is flexed in position to strike the enemy. The left upper limb is stretched horizontally in preparing for enemies on the opposite side. This posture might be taken when the soldier had enemies in front and back. The left, or skull, lower limb is extended; behind this extension there is potential flexion to be produced by the neck reflex.

In contrast, the soldier in Figure App. 6 has his enemies in front and does not have to worry about his back.

Statue of Kagami-jishi: The statue of Kagami-jishi by Hiragushi Denchū is analyzed from the standpoint of the neck reflex as follows. Figure App. 9 shows the posture of a protagonist at one instant. This posture is said to indicate strong movement at this instant. Imaizumi said the following about this statue:

> In 1937, when this Kagami-jishi was performed at the Kabukiza theater, the sculptor went to the theater every evening for 25 days in order to observe the stage from various angles. He talked to Kikugorō Onoe, the main actor of Kagami-jishi, and they decided on this posture, which is most beautiful and is filled with tension. The main actor or his apprentice posed for the sculptor many times even after the performance was over. In order to insure accuracy of depiction,

Fig. App. 9   Fig. App. 10

the sculptor even made a statue of the naked actor (Fig. App. 10). Since Kabuki costumes fit loosely, it was necessary to know the state of the body under the costume. The sculpture is that of the lion at the moment it has stopped after it runs out onto the stage. It has beauty of flow and gives a sense of speed. This is the posture of the moment when one stops suddenly before a very deep valley in order to look down. Active movement is contained in this instant.

All this means that the statue gives a sense of rapid movement. How is this explained in terms of postural reflex?

The lion has come out on the runway very fast, and after exerting rapid forward acceleration it stops suddenly. The naked statue is expressing this forward acceleration by means of the extended upper and lower limbs on the left side, that is, the occipital side.

As mentioned in Chapter 1, Section 4, when the posture that conforms with the neck reflex receives acceleration in one direction, the upper and lower limbs on the occipital side, which have been flexed, become extended. This was exemplified in the three postures of a baseball player catching a ball (Fig. 1.32; p. 28). The same was true for postures in fencing (Figs. 1.28 and 1.29; p. 26) and tennis (Figs. 1.30 and 1.31; p. 27). By the same token, in the statue of Kagami-jishi, the extended upper and lower limbs on the occipital side express forward acceleration. In other words, fast forward movement is depicted in this statue.

The upper limb of this statue on the frontal side is extended as expected from the neck reflex, but the lower limb is flexed. When the lion stops suddenly to look down, this lower limb is flexed, which is opposite to the posture in the neck reflex. This gives force to stopping the body which has been accelerated. All these descriptions may sound like arbitrary interpretations, but I believe that the extended upper and lower limbs on the occipital side in this picture capture the linear acceleration exerted.

Appendix C

Postures of Momentary Standstill in Motion
Neurophysiological Understanding of *Ma* in Japanese Art

Momentary and well-balanced pauses in motion between gravity and anti-gravity forces are seen in sports and in the stylistic poses of Kabuki and other theatrical genres. I have from time to time emphasized that the physiological investigation of these styles and postures is helpful in

Fig. App. 11
Reproduced from a TV commercial for Takeda Chemical Industries, Ltd.

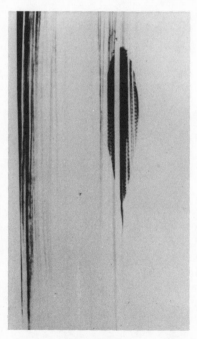

Fig. App. 12
Part of the *Ryūmon-zu*. Courtesy of Daijōji temple, Hyogo, Japan.

301

understanding the meaning of a momentary standstill amidst motion, i.e., *ma*, which is one of the concepts of beauty in Japanese art.

The Masai tribesman on the left in Figure App. 11 has been photographed at the peak of a beautiful leap, in which equilibrium is found between jumping forces and gravity. In this instant of perfect balance we can feel *ma*, a momentary state of weightless standstill. In contrast, the Japanese on the right is stiff and awkward; his jump is not high enough as it is not executed at top capacity: in a word, it lacks *ma*.

When I saw the TV commercial from which the photo in Figure App. 11 is taken, I was reminded of a *sumi* drawing by Maruyama Ōkyo, an 18th-century painter, which shows a carp leaping over a waterfall (Fig. App. 12). The carp is drawn at the peak of its muscular tension. Is it about to jump clear over the waterfall, or will it lose its strength and fall back? Ōkyo's is a beautiful picture of *ma*—rest amidst motion—capturing the instant of equilibrium between the falling water and the ascending fish.

Bibliography

Aronson, L. (1933). Conduction of labyrinthine impulses to the cortex. *J. Nerv. Ment. Dis.*, **78**: 250–259.

Arslan, M. (1953). In: Fortschritte der Hals-, Nasen- und Ohrenheilkunde. I. S. Karger AG, Basel.

Bárány, R. (1907a). Physiologie und Pathologie der Bogengangsapparates. Deutch. otol. Gesellsch., Jena.

Bárány, R. (1907b). Untersuchungen über den vom Vestibularapparat des Ohres reflektorische ausgelösten rhythmischen Nystagmus und seine Begleiterscheïnung. *Monatschr. f. Ohrenh.*, **91**: 477–526.

Basler, A. (1929). Zur Physiologie des Hockens. *Z. Biol.*, **88**: 524.

Braune, W. and Fischer, O. (1895). Der Gang des Menschen. Abh., d. Mathem. phys. Klasse, d. Kgl. Sächs ges. d. Wiss.

Brookhart, J. M. (1979). Convergence on an understanding of motor control. In: Posture and Movement (ed. R. E. Talbott and D. R. Humphrey), pp. 295–303. Raven Press, New York.

Buys, E. and Ryland, P. (1939). Rotation. *Arch. int. physiol.*, **49**: 101.

Doege, R. (1923a). Habitulation to rotation. *J. Exp. Psychol.*, **6**: 1.

Doege, R. (1923b). Thresholds of rotation. *J. Exp. Psychol.*, **6**: 107.

van Egmond, A. A. J., *et al.* (1952). The function of the vestibular organ. *Pract. ORL* (Basel), **14** (Suppl.): 1–109.

van Egmond, A. A. J. (1953). Standardisation et simplification de l'examen de l'appareil vestibulaire. II. *Acta Otolaryngol.*, **43**: 283–289.

Fischer, M. H. (1930). Funktion der Vestibularapparates bei Fischen. *Bethes Handb. Physiol.*, **XV**: 797.

Fischer, M. H. and Babcock, H. L. (1919). The reliability of the nystagmus test. *J. Am. Med. Assoc.*, **72**: 779–780.

Foerster, O. (1932). Sensible corticale Felder. *Handb. d. Neurol.*, **6**: 382–389.

Fukuda, T. (1957). Stato-kinetic Reflexes in Equilibrium and Movement (in Japanese). Igaku Shoin, Tokyo.

Fukuda, T. (1958). Static and kinetic labyrinthine reflex. Functional development of labyrinthine function with rotatory training. *Acta Otolaryngol.*, **49**: 467.

Fukuda, T. (1959a). Vertical writing with eyes covered. A new test of vestibulospinal reaction. *Acta Otolaryngol.*, **50**: 26.

Fukuda, T. (1959b). The stepping test. Two phases of the labyrinthine reflex. *Acta Otolaryngol.*, **50**: 95.

Fukuda, T. (1963). Physiology of training. *Acta Otolaryngol.*, **56**: 239.

Fukuda, T. (1979). On human dynamic postures. *Agressologie*, **20**, B: 99–104.

Gerebtzoff, M. A. (1940). Recherches sur la projection corticale du labyrinthe des effects de la stimulation labyrinthique sur l'activite electrique de lécorce cerebrale. *Arch. n. internat. de physiol.*, **50**: 59–99.

Gesell, A. and Amatruda, C. S. (1947). Developmental Diagnosis. 2nd ed. Hoeber, London.

Goldstein, K. and Riese, W. (1923). Über induzierte Veränderungen des Tonus. *Klin. Wchnschr.*, **2**: 1201–1206.

Grahe, K. (1927). Bei wiederholten Umdrehungsserien nimmt die Dauer des Drehnystagmus. *Handb. d. normal. u. path. physiol.*, **11**: 933.

Griffith, C. R. (1920a). The decrease of after-nystagmus during repeated rotation. *Laryngoscope*, **30**: 129.

Griffith, C. R. (1920b). Concerning the effect of repeated rotation upon nystagmus. *Laryngoscope*, **30**: 22.

Gutner, L. B., Gould, W. J. and Hanley, J. S. (1955). Effect of meclizine hydrochloride (bonamine) upon vestibular function. *Arch. Otolaryngol.*, **62**: 497–503.

Güttich, A. (1920). Die Gewöhnung an Umdrehung. *Anat. u. Ther. d. O.N.H.*, **12**: 57.

Güttich, A. (1940). Über den Antagonismus der Hals- und Bogengangsreflexe. *Arch. Ohren-usw. Heilk.*, **147**: 1.

Güttich, A. (1942). Musklerreflex nach Schwerpunktverlagerung und griechische Plastik. *Arch. Ohren-usw. Heilk.*, **151**: 287.

Harrison, M. S. (1955). Some notes on the practical value of the caloric test in clinical practice. *Arch. Otolaryngol.*, **62**: 459.

Hellebrandt, F. A. (1938). Standing as a geotropic reflex. *Am. J. Physiol.*, **121**: 471.

Hellebrandt, F. A., et al. (1956). Tonic neck reflexes in exercises of stress in man. *Am. J. Phys. Med.*, **35**: 144.

Hirsch, C. (1940). A new labyrinthine reaction. The waltzing test. *Ann. Otol.*, **49**: 232.

Hoff, H. and Schilder, P. (1927). Die Lagereflexe des Menschen. Julius Springer, Wien.

Hoshino, T. and Fukuda, T. (1954). On the postrotatory nystagmus of school boys trained with special rotatory exercises. *Kyoto Zibi Rinsho* (in Japanese), **47**: 769.

Howorth, B. (1946). Dynamic posture. *J. Am. Med. Assoc.*, **131**: 1398–1404.

Jongkees, L. B. W. (1953). In: Fortschritte der Hals-, Nasen- und Ohrenheilkunde. I. S. Karger, Basel.

Joseph, J. and Nightingale, A. (1952). Electromyography of muscles of posture. Leg muscles in males. *J. Physiol.*, **117**: 484–491.

Keele, S. (in press). Understanding human motor behavior.

Kempinsky, W. H. (1951). Cortical projection of vestibular and facial nerves in cat. *J. Neurophysiol.*, **14**: 203–210.

de Kleijn, A. and Versleegh, E. (1930). Experimentelle Untersuchungen über den sogenannten Lagenystagmus während akuter Alkoholvergiftung beim Kaninchen. *Acta Otolaryngol.*, **14**: 356.

Kobrak, F. (1950). Eine unkomplizierte neue Drehprüfung "Der Entschleunigungstest". *Arch. f. Ohren-usw. Heilk.*, **158**: 304–312.

Kornmüller, A. E. (1937). Die bioelectrischen Erscheinungen der Hirnrindenfelder. Georg Thieme, Leipzig.

Leitenstorfer (1897). Das militärische Training auf physiologischer und praktischer Grundlage. F. Enke, Sttutgart.

Luhan, J. A. (1932). Some postural reflexes in man. *Arch. Neurol. Psychiat.*, **28**: 649.

Mach, E. (1875). Grundlinie der Lehre von der Bewegungsempfindung. Engelmann, Leipzig.

Magnus, R. (1924). Körperstellung. Julius Springer, Berlin.

Maxwell, S. S., Burke, U. L. and Reston, C. (1922). The effect of repeated rotation on the duration of after-nystagmus in the rabbit. *Am. J. Physiol.*, **58**: 432–433.

Mittermaier, R. (1954). Über systematische nystagmographische Untersuchungen des kalorischen und rotatorischen Nystagmus. *Acta Otolaryngol.*, **44**: 574.

Montandon, A., Monnier, M. and Russbach, A. (1955). A new technique of vestibular rotatory stimulation and of electrical recording of nystagmus in man. *Ann. Otol.*, **64**: 701–717.

Narita, T. (1954). An acquisition of aireal restistancy of the equilibrium apparatus by means of special gymnastic exercises. *Kyoto Zibi Rinsho* (in Japanese), **47**: 859.

Nishihata, T. (1938). An introduction to otorhinolaryngology (in Japanese). Homeido, Tokyo.

Ohm, J. (1924). Die Entstehung des Augenzitterns. *Neurol. d. Ohr.*, **1**: 1089.

Pacella, B. L. and Barrera, S. E. (1940). Postural reflexes and grasp phenomena in infants. *J. Neurophysiol.*, **3**: 213–218.

Poljak, S. (1932). The main afferent fiber systems of the cerebral cortex in primates. In: Univ. California Publ. Anat. 2.

Rademaker, L. (1931). Das Stehen. Springer, Berlin.

Rossberg, G. (1955). Zur Vestibularisuntersuchung mit ruckartigen rotatorischen Reizen. *Arch. f. Ohren-usw. Heilk.*, **168**: 19.

Schierbeeck, P. (1953). Vestibular Reactions in pigeons after small, well-defined rotatory stimuli, before and after operations on the labyrinth. *Pract. ORL* (Basel), **15**: 87–98.

Sherrington, C. (1920). The integrative activity of the nervous system.

Spiegel, E. A. (1932). The cortical center of the labyrinth. *J. Nerv. Ment. Dis.*, **77**: 504–512.

Spiegel, E. A. (1934). The electroencephalogram of the cortex in stimulation of the labyrinth. *Arch. Neurol. Psychiat.*, **31**: 469–483.

Spillane, J. D. (1975). An Atlas of Clinical Neurology. 2nd ed. Oxford University Press, Oxford.

Steinhausen, W. (1925). Cupulabewegungen beim Frosch. *Klin. Wchnschr.*, **4**: 853.

Tokizane, T., *et al.* (1951). Electromyographic studies on tonic neck, lumbar and labyrinthine reflexes in normal persons. *Jpn. J. Physiol.*, **2**: 2.

Unterberger, S. (1938). Neue objective registrierbare Vestibulariskörperdrehreaktion, erhalten durch Treten auf der Stelle. Der "Tretversuch". *Arch. Ohren-usw. Heilk.*, **145**: 478.

Veits, C. and Kozel, W. (1930). *Lehre v. Kalor. Nyst. Ohr.*, **64**: 521.

Vierordt, H. (1881). Das Gehen des Menschen in gesunden und kranken Zuständen nach selbstregistrierten Methoden dargestellt. Tübingen.

Walsh, E. G. (1957). Physiology of the Nervous System. Longmans and Green, London, New York, Toront.

Wartenberg, R. (1947). The Babinski reflex after fifty years. *J. Am. Med. Assoc.*, **135**: 163.

Winkler, C. (1921). Le System du N. Octavus. *Opera Omnia*, **7**.

de Wit, G. (1953). Seasickness. *Acta Otolaryngol.*, **56** (Suppl.): 108.

Wittmaack, K. (1911). Diagnose spezieller Erkrankungsprozesse im Bereich des Bogengangsapparates und seiner höheren Bahnen. *München Med. Wchnschr.*, **58**: 824.

Wodak, E. and Fischer, M. H. (1923). Experimental Beitrage zu der vestibulare Tonusreaktion. *Z. f. HNO*, **6**: 229.

Index

307